GUIDE TO TEQUILA

GUIDE TO TEQUILA

BY LAURENCE KRETCHMER

PHOTOGRAPHS BY

LAURENCE KRETCHMER AND

ZEVA OELBAUM

Black Dog
& Leventhal
Publishers

Published by
BLACK DOG & LEVENTHAL PUBLISHERS, INC.
151 West 19th Street
New York, NY 10011

Distributed by
WORKMAN PUBLISHING COMPANY
708 Broadway
New York, NY 10003

Manufactured in the United States of America

ISBN: 1-57912-010-5

h g f e d c b a

BOOK DESIGN BY JONETTE JAKOBSON

Library of Congress Cataloging-in-Publication Data

Kretchmer, Laurence, 1965–
Mesa Grill guide to tequila / by Laurence Kretchmer: photographs
by Laurence Kretchmer and Zeva Oelbaum.
p. cm.
ISBN 1-57912-010-5
1. Tequila. I. Title.
TP607 .T46K74 1998
663' .5—dc21
98–9453
CIP

ACKNOWLEDGMENTS

A number of people should be thanked for contributing directly to this book, or indirectly to making it possible for me to write the book. Pamela, for everything, including figuring out the delicate balance between editor and wife. My parents, Dorothy and Jerry Kretchmer, for encouraging me endlessly. My partner Bobby Flay for continuously teaching me to appreciate taste and flavor. My partner Jeff Bliss, who for my entire adult life has pushed me to be ever more critical and ever more analytical. My managers at Mesa Grill: Craig Petroff, Rick Pitcher, and Fran Bernfeld. Craig initially inspired my tequila education and has the most discriminating tequila palate that I know of; Rick is discriminating in general. My managers at Mesa City: Peter Mendelsohn and J.P. François, gaining fast on the Grill. My managers at Bolo: Scott Henkle and Marie Oppenheim, who make Bolo the gem that it is, while allowing me to pursue things like this book. Stephanie Banyas for always making life a little easier and for the awesome food recipes. Billy Steel, an original Mesa Grill bartender who will soon make his one millionth Mesa Margarita as great as his first. Wayne Brachman for a long Southwestern dessert history even with tequila. Manny Gatdula, without whom, my businesses and I would be in a perpetual state of disorder. All of the members of the Mesa Grill tequila tasting panel including Pam; Bobby; Craig; J.P.; Scott; Wayne; Billy; and Steven Kault, chef-owner of Spartina Restaurant; John Gray, manager of the Grand Havana Room; Katie Brown of Lifetime Television; Rick Weisfeld of Bronx Builders; and Daniel Lerner, wine consultant and single-malt Scotch author. J.P. Leventhal for wanting to publish this book and Jonette Jakobson for her design. All of the representatives of the tequila companies in Mexico and in the United States who either received me graciously at their facilities or provided me with information as I needed it. Thanks Mitch. Thanks to Bob Denton, importer of two of the greatest "real" tequilas, El Tesoro and Chinaco—he told me things about tequila that no one else would. Pamela Hunter for the photographs from Oaxaca.

Finally, I owe a unique expression of gratitude to Antonio Ruiz Camarena of the Tequila Sauza Company in Guadalajara. This book would have been far less interesting if Antonio had not been there to help me travel throughout Jalisco and communicate in a language where I possess no proficiency.

TABLE OF CONTENTS

INTRODUCTION

Tequila is raging! No, not raging like a fraternity party during pledge time or a south Florida beach during spring break. Tequila is raging because it has finally moved from its long entrenched niche in the past to its present, rightful place alongside the other great spirits from around the world. Tequilas are now easing their way into the liquor cabinets of the most discriminating drinkers. And they are now found on the top shelves of the back bars of great restaurants where they are being matched, like wines, with different foods on the basis of their various nuances and characteristics.

Despite tequila's new, higher profile, it is still probably the most underappreciated and misunderstood of the world's great spirits, with its history and use shrouded in deep-seated misconceptions. In fact, it has been my experience that most drinkers lack even a fundamental understanding of the defining characteristics, classifications, and different quality levels of the various tequila types.

At our restaurants Mesa Grill and Mesa City in New York City, tequila is by far the most popular spirit. Our customers drink tequila straight and on the rocks, they shoot tequila, they sip tequila and drink tequila mixed in countless Margarita variations. Most importantly, they enjoy tequila by itself or with our unique style of boldly flavored American Southwestern cooking. And while the increasing popularity of Southwestern as well as Mexican cuisines has definitely contributed to the increasing popularity of tequila—due, no doubt, to their common heritage—it is the extensive range of the types and the qualities of tequila, together with the depth of quality, that has changed the most.

While tequila continues to be one of the world's most popular spirits—Americans alone consume over five million cases a year—it is tequila's image and place in the world of food and beverage that has changed drastically. Handcrafted, 100 percent blue agave tequilas, some young and fresh and some aged, are surging in popularity, right along with single malt Scotches and small batch bourbons. Ironically, the vast majority of tequila consumed in the United States is still not of the most authentic, highest quality brands, but rather is the blended or *mixto* white and gold varieties that continue to fill Margarita and shot glasses in staggering numbers. Furthermore, there is rarely a correlation between the price of a tequila, the care taken in its production, and the quality of the resulting product.

Knowledge, as the saying goes, is power, and a better understanding of not only the history of tequila but of the various ways it is made today, should smooth the transition from the familiar shot glass, to the snifter that fine tequilas deserve.

My hope in writing this book is to provide a basic orientation in the "whats," "whys," and most importantly, the "hows" of the ever increasing variations of tequila that are becoming available, including descriptions of how they are made and how they may be drunk and most appreciated. I have tried to draw a road map which can be used in the search for the true essence of the blue agave. Armed with this information, one should be able to learn how to discern between a tequila that looks good and a tequila that tastes good. If my objective is met, then I will have not only helped more people to enjoy tequila, but I will have helped people to enjoy tequila more!

L. K.
New York City
1998

The mural by Gabriel Flores at the Sauza factory in Jalisco depicts the legend of tequila's creation, the traditional tequila production process and the various effects of tequila on its drinkers.

WHAT IS TEQUILA?

THE LEGEND OF TEQUILA'S CREATION

According to an old tale, during Mexico's pre-Hispanic times, tequila was discovered when a bolt of lightning struck an agave field. The bolt tore into the heart of one of the plants and the heat of the lightning bolt was so hot that it burned the heart of the plant for several seconds, causing the plant to become not only cooked, but also naturally fermented. The shocked natives noticed an aromatic nectar coming out of the plant. With a certain amount of fear, but also with reverence, they drank the nectar which they deemed to be a miraculous gift from their gods. It was as if the lightning's fire had turned into a new and mysterious drink which they named *vino mezcal,* the mezcal wine.

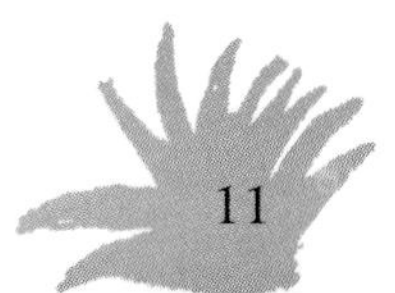

WHAT IS TEQUILA?

There are probably more myths surrounding tequila's origin and its attendant conventions than most other beverages. But before dispelling the myths of cactus juice and worms swimming in the bottoms of bottles, let's start by finding out what tequila is. In this effort we get a lot of help from the Mexican government, which has outlined a strict set of standards according to which all tequila must be made if it is in fact to be called tequila.

Tequila is a distilled spirit that is made only from a plant whose technical name is the *Agave Tequilana*

The blue agave, or agave azul is one of nearly 400 species of agave now categorized in its own plant family—the Agavaceae. Definitely not a cactus.

Weber, blue variety. The more common name is shortened to the "blue agave," and that is how we refer to it in this book. While nearly 400 different varieties of agave plants, also known in Mexico as *maguey* (mah-gay), have been identified by botanists, it is the blue agave species, often referred to as *agave azul,* which Mexicans call "the plant of the gods." Many believe the blue agave is from the cactus family. It is not. In fact, the agave plant, which is classified not as a fruit but as a succulent, used to be classified in the same family as the Lilys and Aloes families (kind of like a cousin), but it is now classified in its own family, the Agavaceae. The blue agave's "plant of the gods" alias is the result of its singular importance in the making of a vital economic commodity. Some of the other varieties or species of agave are used to make other beverages, like pulque and mezcal, and are also used in

Throughout the Amatitán Valley, fields of blue agave color the countryside in shades of azure. At Tequila Sauza's Rancho El Indio farm, agave fields welcome visitors as the infamous "Tequila Mountain" looms in the background.

the production of shampoos and creams as well as certain industrial products.

Much like cognac, which is a brandy, scotch, which is a whisky, or champagne, which is a sparkling wine, tequila, which is technically a mezcal, is a beverage of a particular place. Not only must it be produced in Mexico, but more specifically, it can only be produced in one of five approved regions of Mexico: the entire state of Jalisco and certain villages in the states of Nayarit, Tamaulipas, Mihoacán and Guanajuato. A beverage made similarly to tequila, but not in one of the prescribed locales, may be called a mezcal. To use the brandy analogy, cognac is a special type of brandy made a certain way, from certain types of grapes, and only in a prescribed region within France. So all cognac is brandy but not all brandy is cognac. In much the same vein, tequila is a special type of mezcal, made in a specific way, from a single variety of the agave, only in specified areas of Mexico. So while all tequilas are technically mezcals, not all mezcals are tequilas.

In actuality, most tequila is produced in Jalisco in one of two concentrated areas near Guadalajara, in or near the town of Tequila. As I will explain in a later chapter, some grades or classifications of tequila—the lesser-quality mass-produced ones—may actually be bottled outside of Mexico. However, the growing, farming, fer-

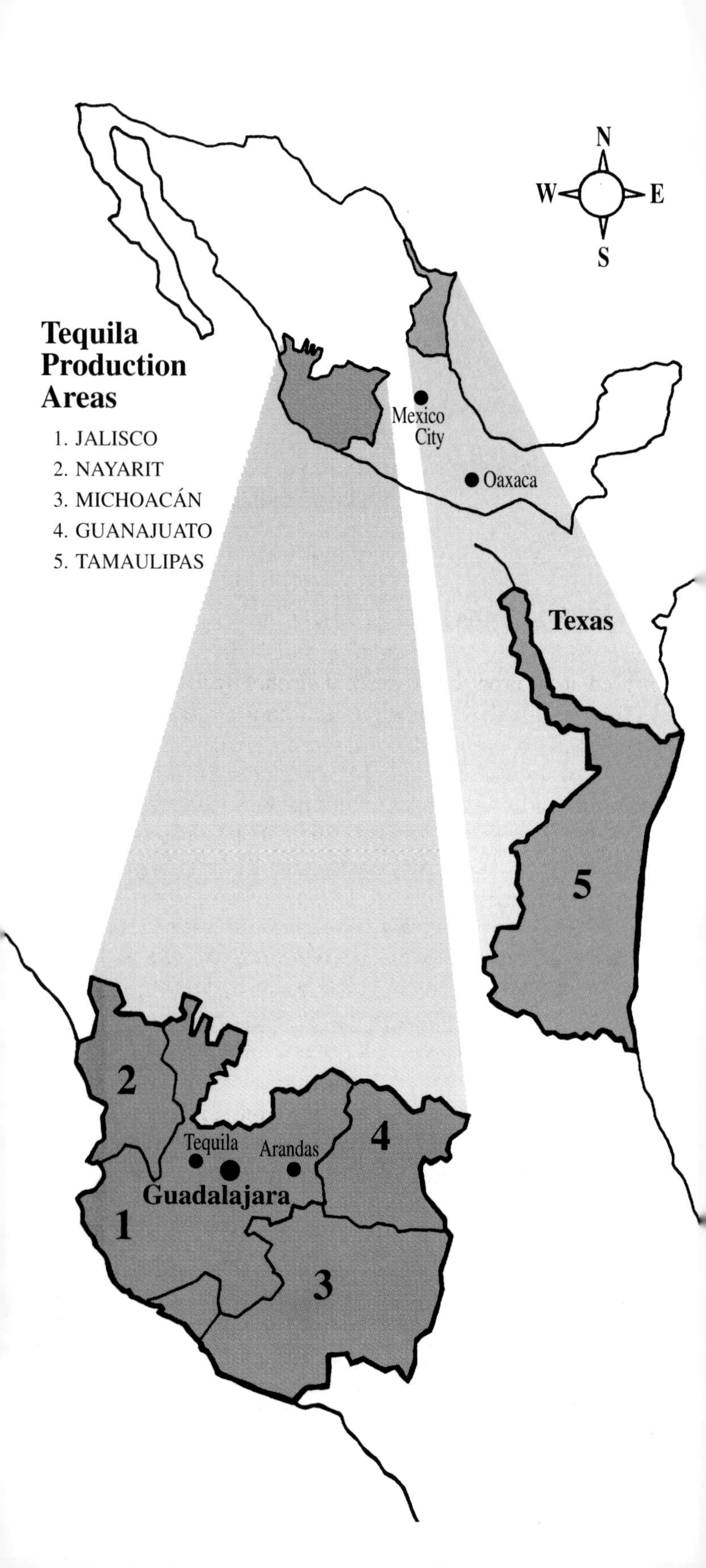
Tequila Production Areas
1. JALISCO
2. NAYARIT
3. MICHOACÁN
4. GUANAJUATO
5. TAMAULIPAS
N
W
E
S
Mexico City
Oaxaca
Texas
Tequila
Arandas
Guadalajara
1
2
3
4
5

menting and distilling, which are the most important, defining stages in tequila production, must all take place within the approved regions. Furthermore, for the "premium" grades of tequila, even the bottling must be done in Mexico.

The final basic criteria for tequila authenticity concerns the ingredients used in its production. In order for a product to be called tequila, it must contain at least 51 percent fermented juices from the blue agave plant. The other 49 percent may consist of various sugars that are added during the fermentation process.

Despite myths to the contrary, tequila is no more "potent" than other distilled spirits. In reality, tequila meets the same standards of proof, or alcohol content, as do other distilled spirits. While various tequila producers distill their products to different levels of alcohol content, for consumption in the United States—and just like gin, vodka or rum—after dilution tequila, is eventually bottled at 80 proof or 40 percent alcohol by volume.

There are further regulations governing tequila production with respect to the aging of the product. Those requirements will also be explained later in the discussion of the different classifications of the product. For now, it should be understood that tequila production is regulated by the Mexican government. All of the regulations, including allowable areas of production, allowable methods of production, restrictions with respect to aging and even labeling, as well as other requirements which define the different types or classifications of tequila, are defined in a set of laws called the *Norma Oficial.*

According to the *Norma,* every bottle of tequila is labeled with a "NOM," which is a number assigned by the government indicating where the tequila in a particular bottle was produced. Presently there are approximately fifty different government-approved distilleries legally producing tequila, each with its own individual number. Often, the name of the distillery or the company that owns the distillery will be different from the brand of tequila itself. More on that in the next section.

On the left: ***The shaded areas indicate the regions where tequila may be produced in Mexico. Jalisco and Tamaulipas are the only two states currently producing tequila.***

Every label on a bottle of tequila must include a "NOM" number identifying exactly which factory manufactured the tequila contained inside that bottle.

Now that the scientific, technical and legal aspects of tequila production have been discussed, where did the cactus juice and worm myths come from?

While most myths can be explained by stories that are rooted in reality, in the case of the cactus juice theory, I can only guess that somebody who consumed a bit too much tequila wandered into an agave field in the dark of night, stuck himself on the sharp leaves of an agave plant and confused that plant with a cactus. While there are several other alcoholic beverages besides tequila which are made in Mexico from other species of agave, none of them are actually made from the fermented juice of a cactus.

As for the worm myth, I have seen hundreds of different tequilas, but never once have I seen a worm in a tequila bottle. There never was, and probably never will be, a worm in a legitimate bottle of tequila. The confusion here is with mezcal, tequila's forerunner, which is a distilled spirit made in areas not designated for tequila production, from varieties of agave other than the blue agave variety used to make tequila. Mezcal differs from tequila not only in the way it is made, but also in its taste. While most of the best brands of mezcal being produced today are wormless, there have been, and still are, examples of the inclusion of worms in the bottling of mezcal. But remember, there never was, and probably never will be, a worm in a legitimate bottle of tequila.

THE HISTORY OF TEQUILA

TEQUILA'S ORIGIN

Mention tequila, and the stories start flowing. Undoubtedly there will be the ones about dancing on bars or kissing strangers, and some know-it-all will proclaim that there is something in tequila that makes you a crazier, nuttier, wackier sort of drunk than any other spirit does. These are just a few of the many myths surrounding tequila. In fact, it has never been proven that tequila has a physical effect that is any different from other alcoholic beverages. (In that regard, booze is pretty much booze—its how much you drink and how fast you drink it that makes the difference.)

The roots of tequila's mystique lie in its rich history, beginning with the Aztecs, an ancient civilization known—among other more noble accomplishments—for its warring ways and its taste for human sacrifice. The Aztecs found a variety of uses for the plant they called *metl.* From its barbed, fibrous spears and pineapple-like heart they made thread, rope, clothing and shoes—even paper and brushes, since they were a literate society. From its "sap" they made *pulque,* the crude fermented (but not distilled) brew that is the precursor to tequila (which is still consumed in Mexico to this day, nearly 1,000 years later).

The *metl* was considered a gift from the gods by the Aztecs, and *pulque* was their very blood. In order to insure a steady supply of *pulque* from the gods, human blood was sacrificed in gruesome ceremonies. And since *pulque* was highly valued for what the Aztecs believed to be its curative qualities as well as its hallucinogenic and relaxing effects, consumption of *pulque* was reserved for ritualistic use of priests, nobles and the infirm. Severe penalties were enforced for misuse of *pulque* to prevent the widespread public drunkenness that might otherwise have resulted. Average Aztecs, however, were allowed to drink themselves silly for a mere five days a year during what was called the "Days of the Dead" at the end of their calendar year.

When the Spaniards first arrived in Mexico in the early 1500s, they were unimpressed with the low alcoholic strength, sour taste, odd consistency and perishability of the drink that the native Indians worshipped. The process of distillation was unknown to the natives, but old news to the Spaniards, who were accustomed to the brandies, wines and rums of their homeland, which were now unavailable to them. Necessity is the mother of invention, and it was not long before they tried applying distillation to the sap of the plant from which the natives fermented *pulque.* There were many varieties of the plant which they called the *maguey* (after a similar-looking plant they knew from the Caribbean), and the Spaniards experimented with different varieties and distillation techniques until they hit upon something more to their liking. They called this new distilled spirit *vino mezcal.*

The Spaniards in Mexico eventually established European trade and went back to drinking their brandy, wine and rum, but not before teaching the Mexicans to distill this mezcal wine. The natives acquired a taste for mezcal and began producing it from different varieties of agave, which grew all over the country, including a little agricultural town called Tequila in a valley bordered by a once active, 3,000-foot volcano in southwestern Mexico. Most think the name tequila comes from the Indian word *tequitl* for work, job, or task, or possibly from the verb *tequi*, which means to cut, work, or toil. And cut and toil they did in Tequila, where a particular variety of the *maguey* grew like a weed in the rich, volcanic soil.

In the 1600s, the production of mezcal was still on a very small scale. The town of Tequila became known for its outstanding mezcal, and as demand for it increased and the population grew, so did mezcal's production. As it grew, the Spanish government began regulating its production and imposing taxes, as governments will. Perhaps the popularity of mezcal grew a little too much for the liking of the Spanish crown, because the production of mezcal was forbidden by them in 1785 in the hope that it would increase the importation of their own wines and spirits, thus helping the economy back home. The mezcal trade was forced underground until a new king, Ferdinand IV, lifted the ban in 1795 and granted the very first official license to produce mezcal wine to Don Jose Maria Guadalupe de Cuervo. Does that name sound vaguely familiar?

The nineteenth century in Mexico was marked by war:

A street in the town of Tequila, Jalisco, where many of the largest tequila factories are located.

the eleven year battle for Mexican independence from Spain; The War of Reform—a civil war lasting for three years; and another European invasion, this time from the French. It was at some point during this period that the *maguey* began to be cultivated in and around Tequila as a crop. Eventually, a scientific classification was given to all of the maguey plants—the genus Agave. The *Agave Tequilana Weber* or *Blue Agave* was the variety of *maguey* from which mezcal was made in Tequila, and the mezcal wine of the region took on the name of its source.

The first distillery to export tequila to the United States, in or around 1873, was owned by Don Cenobia Sauza, another familiar name. By the turn of the twentieth century, exportation grew as mechanization and scientific advances took tequila production and transportation into the modern age. Prohibition probably gave tequila's popularity in the United States a kick, since it was easily smuggled over the border. Officially recorded exports stayed at a minimum, though, until just before World War II, when the American market really began its love affair with the spirit of Mexico due to a halt in the import of whisky from Europe. And tequila was exported to the United States in another form that also served to fuel the American market for tequila—and supply it with the imagery that still lends tequila much of its wild "Ranchero" associations today—the popular Mexican films of the '30s and '40s. From a minuscule 6,000 gallons in 1940, tequila exports had exploded to 1.2 million gallons by 1945, with the United States by far the largest consumer.

In 1948, by the end of the war, the bottom had fallen out of the American tequila market to an all-time low of 2,500 gallons. Within Mexico, however, tequila consumption and production continued to grow as the native drink became identified with Mexican pride in their culture and heritage.

No other factor has had as great an effect on the tequila industry as the growth in popularity of the Margarita cocktail in the United States. Invented sometime during the 1930s or 1940s, there was most likely a woman named Margarita with a taste for tequila who caught a creative bartender's eye. (There are even more stories alleging to recount the invention of the Margarita than there are about the invention of the Martini.) The

Margarita really hit its stride in the United States in the 1970s, and has been the number one most popular cocktail in the country ever since.

But a funny thing happened to tequila on its way to popularity. The need for more and more product for export meant that producers started looking toward mechanization and shortcuts to keep supply equal with demand. The blue agave plant takes eight to twelve years to fully mature, and the process of extracting its valuable sugars for fermentation is an arduous one. There were many less expensive sources for sugars and modern means of processing now available to tequila producers, and naturally, they took advantage of them.

In an effort to preserve their national treasure, in 1978 the Mexican government established *Normas* to govern standards of production, quality and labeling. Functioning in much the same way that the French Appellation Contrôllé maintains a set of standards for wine production and labeling, these *Normas* defined where and how tequila must be made in order to have the right to call itself tequila. At the same time, strict standards were set for the various classifications of tequila that were developing.

HOW IS TEQUILA MADE?

An understanding of the varied processes used to produce tequila will lead to a greater understanding of the product itself. On the way from the fields where agave is grown, to the bottle from which it is consumed, the producer makes a number of choices at each stage of production that in turn affect the way the product tastes. One would assume that the larger the operation, the more automated the production process becomes. However, my study and observation of various tequila producers' operations show that, with an eye on quality control, there are some producers who, despite their size, have maintained more of the old-fashioned methods than one might expect.

GROWING AND HARVESTING

With over five hundred million kilos of agave being used in tequila production in a single year, and over one

A jimador at work harvesting the agave plant.

hundred thousand acres of blue agave being grown, the growing and harvesting by hand of the agave plant requires far more manpower than any other step in the production process.

Some producers grow all their own agave, some buy all their agave from other growers, and some producers both grow and buy their agave depending on their production needs. Very few producers use only their own agave or what they like to refer to as "estate-grown" agave, and those that do will make a case for their ability to exert greater control over the harvesting process, and therefore the quality of the resultant product. They also maintain that they can better determine the precise moment to harvest the plants that are perfectly mature. Producers using estate-grown agave can also control the exactness of shearing the leaves from the heart of the plant, thus ensuring a purer, better product.

The tequila production process begins with the *jima,* or harvest, and the *jimador,* who is the harvester of the agave plant. The method of harvesting the blue agave

On display at Sauza's Rancho el Indio are the jimador's various tools of the trade—used over 100 years ago and still used today.

plant does not differ significantly from producer to producer. Despite all of the modern advances in agriculture, very little has changed in the way that the *jimador* works. In fact, the same mostly handmade tools that were used over a hundred years ago are still used today.

An important difference between the production of tequila and many other spirits, even wine, is that the source from which the product comes is used only once. In other words, when an agave plant is harvested or removed from the ground, it is consumed and gone forever. Nothing is left behind except for the hole where the plant lived for its eight to twelve years of life. However, during the agave plant's life cycle, it reproduces, fostering the growth of future agave plants over another eight- to twelve-year period. This process is very different from that of grains used in the production of other spirits, and grapes harvested from vines for use in wine production, which bear their product over and over again.

Using a* talache, *in one motion a* jimador *creates a hole in the ground and "buries the heart."

When an agave plant is in its fourth or fifth year of growth, it produces "sprouts," or offspring, which grow adjacent to the mother plant. The farmer uses a tool called a *barreton* to extract the "sprout"—which is really a "shoot"—from the roots of its mother plant. The sprout is actually one year old when this extraction takes place, and it is done in March, April and/or May when the soil is most dry. June through October is the rainy season and is generally not an ideal time for the replanting of agave.

The farmer then holds the extracted baby plant in his hand, and using a tool called the *machete carto* (short machete), he trims off the dried leaves making the plant smaller and easier to handle and transport. At this point he will also cut off the bottom or base of the plant in order to look at the condition of its heart. If the heart is all white, then the plant is apparently healthy and can be replanted. If there are black spots, then the plant is ill, and it will not be replanted but instead be reused as fertilizer.

The next step in the harvesting of the baby agave plants is the replanting of the shoots or seedlings. The *talache* is the tool used to plant the sprout or "bury the heart." After a tractor prepares the soil to make it workable, the plant shoots are laid one to one-and-a-half meters apart in rows which are between one-and-a-half and three-and-a-half meters apart, depending on the soil's properties. Where the soils are richer in minerals, the shoots can be planted closer together. In

Above: ***A jimador has extracted an agave "sprout" from the ground.***
Below: ***After cutting off the bottom of the plant, he examines the condition of its heart to determine its health and suitability for replanting.***

If allowed to grow, the flower spike, or quiote, will grow to great heights, depleting all of the precious sugar from the heart of the plant.

one motion, the farmer strikes the ground with the *talache,* opening up a hole for the plant, extracts the *alache,* and swiftly places the plant (or heart) in the ground and covers up the base of the plant with some earth. He then pats the earth down with his feet and moves on to the next plant. These baby plants will now be on their way to a full cycle of reproduction and growth on their own, until they are deemed mature enough to be harvested for use in making tequila.

During the plants' growth cycle, the farmers do two important things to ensure proper growth. First, they meticulously clean away the weeds and grass that grow in the fields around the agave plants. If proper care is not taken to maintain these fields, the sun and natural rains are not easily able to feed the plants. The *coña de limpia* is one tool used to clean the weeds and grass away. The *casonga,* very much like a sickle, is another tool used for cleaning in the fields. The second important task is treating the plants with fertilizers and nitrates during the eighth to twelfth years of growing to ensure that diseases do not develop.

After five or six years of growth, the points of the leaves are trimmed to make the heart grow larger. This cutting, using a tool called the *machete barbeo,* is done to ensure that the sugars develop more in the heart rather than in the leaves. The intention is to force the heart to grow plumper since that is essentially the "meat of the plant," where its value is located. In female plants, the *quiote* or "flower spike," may grow straight up out of the center of the plant. This blossoming occurs in the fifth or sixth year. If allowed to grow to its full height, it would dry out the heart of the plant by depleting all of its sugars. To prevent this from occurring, the flower spikes are immediately cut off, forcing the continued development and concentration of sugars in the plant's "heart." The trimmed spikes are often cooked and sold as sugarcanes. The *jimador* carries additional tools during the *jima* the *triangulo,* which is a sharpening tool and the *bule o guaje,* which is a hollowed-out plant used as a water canteen.

After about eight years, the agave plant begins to dry as it matures. The farmers then closely monitor the plants over the next few years, watching and waiting for precisely the best time to harvest them. At that moment, there will be evidence of a greenish-yellowish color inside the leaves close to the heart of the plant. The appearance of that hue signals to the *jimador* that it is harvest time.

Using an extremely sharp tool called the *coa,* the *jimador* begins to shear the *pencas* (spiny leaves) off the plant quickly. Exposed are the heart or central mass of the plant, which looks just like a giant pineapple—thus its popular name, the *piña,* also sometimes called the *cabeza,* or head of the agave. With his *coa,* the *jimador* then digs into the ground, searching for the plant's roots and eventually severing the plant from its roots, which are 30 to 40 centimeters deep. After he works the plant out of the earth, the *jimador* turns it over as he continues to shear the length of the leaves off all the way, until he has a clean heart or *piña.* These harvested plants can be very heavy, ranging in

As soon as the flower spike appears in the female agave, it is cut at its base.

The jimador's most important tool—the coa is sharper than a butcher's knife.

The jimador has selected a mature plant for harvest. Here he begins extracting it from the ground.

The jimador completes the trimming of the cabeza, *or head of the agave, leaving the cleaned* piña *or heart.*

The trimmed piñas are collected from the fields by wheelbarrow before they are trucked off to the various tequila production factories.

weight anywhere from 80 to over 175 pounds.

There are two primary areas in the state of Jalisco where blue agave is grown and tequila is produced. About thirty miles west of Guadalajara, at the foot of the "Tequila Mountain," lies the Amatitán Valley and the town of Tequila. Many of the largest distilleries, including those owned by Tequila Cuervo and Tequila Sauza, are located in the town of Tequila. Nearby, in the towns of Arenal and Amatitán, are several other distilleries, including Tequila Herradura.

The other principal tequila-producing area is located about forty miles east of Guadalajara, near the valley area known as the Highlands, or *Los Altos,* but at an elevation approximately 1,500 feet higher than the valley. In this area, mostly in the towns of Arandas, Atotonlico el Alto and Zapotlanejo, as well as in a few other small villages, there are another twenty or so distilleries with several new facilities just completed or in the building or planning stages. It is in Arandas that the giant manufacturer, Seagram's, has just built a new facility where they

A truck full of piñas *waits outside the gates of a tequila factory to drop off its bounty.*

The "reception area" at Tequila Tapatío, in Arandas.

Before cooking, the larger piñas *are cut into smaller pieces.*

will attempt to keep production up with the great demand for their Patrón brand.

Owing to each region's different soil properties, the size and taste of the *piñas* harvested are also different. The *piñas* harvested in the Lowlands—specifically the Amatitán valley, for instance, are typically smaller than those from *Los Altos.* These smaller *piñas* range in weight from 60 to 90 pounds. In the eastern area Highlands or Los Altos region, however, the soil is richer and rather dark red in color, due to its high iron oxide content, which enhances the plants' sugar production. A sweet character is imparted to all the plants grown in this area, not only the agave but also the corn and strawberries. The richer soils of *Los Altos,* which foster the *piñas'* growth to as heavy as 200 pounds, also permits denser planting. While in *Los Altos,*, adjacent plants in a line are the same 1.2 meters apart, the rows themselves are actually 1.6 meters away from each other rather than the 3.5 meters found in the lower lands of the Tequila valley.

After harvesting, the *piñas* are loaded onto pickup trucks the same day and taken to the various distilleries where the tequila production process begins. In one day, a *jimador* might harvest three to four tons of agave plants, which might be as many as 150 plants. He is paid 35 pesos per ton (a ton equals1,000 kilos), so in one day he might earn 100 pesos.

COOKING AND MILLING

As the trucks loaded with piñas arrive at the tequila factory or *fabrica,* the plants are offloaded in what is commonly referred to as the "reception area." This is where the plants are received and prepared for the production process. Before the *piñas* are cooked, they are cut, usually into halves or quarters, so that they will better cook through.

The cooking of the *piñas* is one area where the different producers vary in their methods.

The traditional method for cooking the *piñas* involves loading them into brick or concrete ovens, or *hornos,* which hold around 50 tons of agave plants. The workers load the *piñas* into the ovens by hand, packing the oven until it is jammed. This is done in order to maximize the

At a smaller tequila factory, a traditional* hornos*, or oven, where the agave is cooked.

At the large Herradura factory, workers have the help of a conveyor belt to load one of the several ovens.

amount of agave that can be cooked at once. After the doors are tightly closed, steam is injected into the oven, in effect creating a pressure-cooking environment. The agave is typically cooked at a temperature of 140° Farenheit for a period ranging between twenty-four and thirty-six hours. Then the agave is allowed to cool for another twenty-four hours before being removed from the oven. During the cooking process, the juices that are released from the plants are known as *agua miel* or honey water. They are collected in the bottom of the ovens and transported to a holding vessel for later use.

A more modern method used at many factories is to cook the *piñas* in an autoclave, which is a large stainless

At La Escondida, a small factory in the town of Arenal, tequila is made by a company called Tequila Parreñita. A worker prepares the oven for cooking the agave by packing the perimeter of the oven door with mud. This seals in the steam injected to cook the agave.

At El Llano, the factory in the town of Tequila that is owned by Destiladora Azteca de Jalisco, yet another type of oven—a horizontal autoclave—is used for cooking agave. This steel cooker is capable of cooking agave in just seven hours.

After twenty-four hours of cooking and twenty-four hours of "resting," the agave takes on a new appearance.

Most of the factories now use a modern milling machine to shred the agave and extract its juices after cooking.

Water washes the sugars off of the shredded agave.

steel tank or vessel that acts like a giant pressure cooker. In some factories these autoclaves are big vertical stainless steel tanks which are loaded from the top and then unloaded at the bottom after the cooking is completed. Some other factories have horizontal steel ovens, which look like tubes lying on their sides. These vessels are loaded from the back forward, and after cooking are unloaded through an open door in the back. Cooking in an autoclave permits high temperatures to be reached quickly and shortens the total cooking and cooling time. The effect on the agave, however, is the same no matter which cooking method is used.

During the cooking, the starches in the plant are converted into sugars which, in the next stage, will be fermented into alcohol. During cooking, the agave turns a dark orange-brown color, and it becomes very fibrous so that it can be pulled apart in strands. The taste of the cooked agave at this point is very reminiscent of cooked yams or sweet potatoes, but it is perhaps even sweeter.

After the agave has been cooked, additional juices must be extracted. This is done in the milling process. Depending on how modern the facility is, the cooked *piñas* are transported by either conveyor belt or by hand to the milling or juicing machine. A modern milling or juicing machine has a series of grinding blades that extract the juice from the cooked fibers as they pass through on a conveyor belt. At the same time, water is sprayed on the fibers, washing the converted sugars off of them. These extracted juices, or *agua miel,* are collected and transported to the next holding vessel.

In a more old-fashioned application, the process of extracting the agave's juices from its fibers is done by hand. Using a *tahona* a huge stone wheel weighing as much as two tons, the agave is crushed in a cobblestoned pit by the continuous circular rolling of the wheel. As this gigantic wheel crushes the agave fibers, juices are released, and they are then removed and taken to the fermentation area. Again, in older methods of production, this transporting was done by hand, while in the newer facilities it is all done by conveyor. To my knowledge, Tequila Tapatío, which is the producer of the El Tesoro brand, is the only distillery still using a *tahona* as its primary means of crushing or milling the agave to extract its juices.

FORMULATION AND FERMENTATION

After all of the juice has been removed from the cooked fibers in the milling process, the resulting product is ready for the fermentation stage where fermentable sugars of the agave juice are converted into distillable alcohol.

It is at this point that a decision is made with respect to whether the juice will be used to produce a 100 percent agave tequila or a blended or *mixto* tequila. If 100 percent agave tequila is to be produced, then the juice will go straight into the fermentation vessels, but in the latter case, an intermediate stop is made at the formulation tanks.

In the production of non-100 percent agave tequilas, the additional sugars are added in the formulation stage. *Piloncillo* (diluted sugarcane or molasses juice) is typically added. For 100 percent agave tequila, only the pure juice of the agave is used with some added yeasts.

During the fermentation stage, the sugars are in effect eaten by the newly introduced yeasts, and they begin their conversion into alcohol. Even when 100 percent agave tequila is made, there is some variation in the particular techniques utilized in the fermentation process. A few of the producers pride themselves on their use of exclusively "natural yeasts." By this, they mean that all of the yeasts utilized are naturally occurring and most often are cultivated and maintained on the factory grounds. At many other distilleries, however, commercial yeasts and other catalyzers are introduced to expedite the process.

Smaller facilities typically utilize fermentation tanks ranging from 8,000 to 10,000 liters, while larger facilities accommodate tanks up to 75,000 liters. With the introduction of chemical catalyzers, which speed the growth and spread of yeasts, the process can be completed in a period of thirty-six to seventy-two hours. This accelerated process is mostly used in the production of non-100 percent agave tequilas. Some producers, however, pride themselves on their "natural" production methods. In this case, they are referring to their use of exclusively natural yeasts and their refusal to use any chemical catalyzers to speed up the fermentation process. In its natural mode, this process can take anywhere from five to ten days. Often, the fermentation tanks will be heated slightly in order to initiate the fermentation process. Once the fermentation process is under way, the juice heats up,

At Tequila Tapatío, where El Tesoro is made, a worker is dumping the cooked agave into the pit of the rare tahona *that is still in use.*

In the formulation stage, yeasts and—in some cases—additional non-agave fermentable sugars are added in the production of mixed or non-100 percent agave tequilas.

In a modern tequila factory, agave juice, or **agua miel**, *is pumped into stainless steel fermentation tanks, which may vary in size from 8,0000 to 75,000 liters.*

Above and below: *Early and late stages of fermentation as the yeasts begin to act on the sugars in the agave juice. As carbon dioxide is released, a reaction occurs.*

Fermentation takes place in small wooden tanks at Tequila Tapatío where the crushed agave fibers remain in "the mix."

At Tapatío the agave juice is introduced into fermentation tanks by hand.

On the left, a pot still arrangement at a smaller, more traditional factory, compared to a more modern setup below, where several stills made of stainless steel for the second stage of distillation are positioned behind the first distillation stills.

carbon dioxide is released and you can actually see the conversion of sugars taking place. The top of the tanks will begin to bubble and "roll" as a light brown, creamy-colored foam develops on the surface of the tank. At this point the production of alcohol has begun, but only to a point of about 5 percent alcohol by volume.

DISTILLATION

In the final stage of the production process, the *mosto*, or fermented juice, is distilled. According to Mexican law all tequila must be distilled twice. From the fermentation tanks, the *mosto* is transported either mechanically or by hand to the still.

In a pot still or column still, the liquid is heated to alcohol's vaporization point where the alcohol vapor is removed from the *mosto* and carried off to a cooling condensor. The alcohol vapor is then cooled and condensed and the resulting product is called *ordinario*, which will be between 20 percent and 30 percent alcohol. The first and last portions of the distillation, known as the "heads and tails," which may contain "bad alcohols," including congeners or impurities are then discarded. The remaining *ordinario*, or "good alcohol," then goes through a second distillation where the tequila is produced.

The product of the second distillation may vary depending on the type of tequila being made. In premium tequila production, the product often will be distilled

Inside of a pot still, a coil is heated up to a temperature high enough to remove alcoholic vapors from the mosto.

At Jose Cuervo's La Rojeña, copper stills are used to bring the alcohol to a very high level before dilution later.

to the precise levels required for the finished product. This is 80 proof or 40 percent alcohol by volume. In the case, however, of nonpremium, mixed or "bulk tequila" production, the second distillation may create a product that is distilled to 55 percent alcohol by volume or 110 proof. These are later diluted with purified water before final bottling and shipping.

AGING

At the completion of the distillation process, the product that comes from the still is tequila in its pure form—which is actually a *blanco,* or white tequila, sometimes also referred to as *plata,* or silver tequila. Depending on the type of tequila being produced, the product may now be either bottled, aged or prepared for bulk shipment to another location where, in the case of the least expensive quality of tequila, it undergoes a final blending with added colors and flavors and is then diluted.

Tequilas that become *reposados* or *añejos* are aged in wood vessels of varying sizes. The wood imparts a color, a range of flavors, and fosters development of a different texture or smoothness and body. *Reposados* are typically aged in large wooden tanks, often redwood, but also sometimes in small oak barrels, while *añejos* are almost always aged in Kentucky bourbon barrels. Varying ages of oak barrels may be used. The newer barrels give off more color and tannins, lending structure to the tequila,

Larger tanks holding up to 30,000 liters, which are typically made of redwood, may be used for aging reposado tequilas. Añejo tequilas must be aged in small oak barrels no larger than 200 liters.

while older barrels give off less color and less tannin and tend to soften the tequila. Because the oak loses some of its effectiveness over time, the use of older barrels creates a smoother product with less color and body. The newer barrels create a stronger product from the effect of the fresh tannins in the wood, yielding a product with more color. With variations from barrel to barrel, the more meticulous tequila producers carefully blend the tequilas from the different barrel lots to create a balance between the various lots' tannins and softness, resulting in wonderful, handcrafted tequilas.

Depending on the type or category of tequila being produced, aging takes place for between two months and four years. Like many other spirits, such as scotches, brandies and fine wines, tequila will evolve substantially during the first four years. After four years, however, unlike those beverages, tequila not only stops improving but may even begin deteriorating as a result of excess exposure to the tannins in the wood where it is being stored. Unlike many other premium spirits, tequila does not improve in its bottle over time, either. So, if you have tequila that you have been saving, the best thing that you can do with it is drink it!

TEQUILA TYPES AND CLASSIFICATIONS

THE DIFFERENT TYPES OF TEQUILAS

Tequila's growing popularity has spurred manufacturers to increase the various types and classifications of tequila they are producing. Depending on the actual ingredients and production methods used to make each of the many products, these numerous varieties can be categorized in ways that will reveal to the drinker a great deal about a product even before it is tasted. The Mexican government has outlined four strict classifications categorizing the various tequilas according to their aging and to a lesser extent their content.

TEQUILA CLASSIFIED BY AGAVE CONTENT

The most basic distinction among tequilas is between

TYPICAL TEQUILA PRODUCTION PROCESS

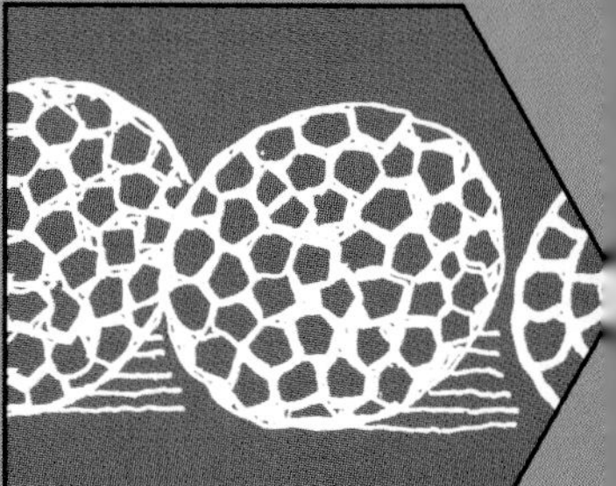

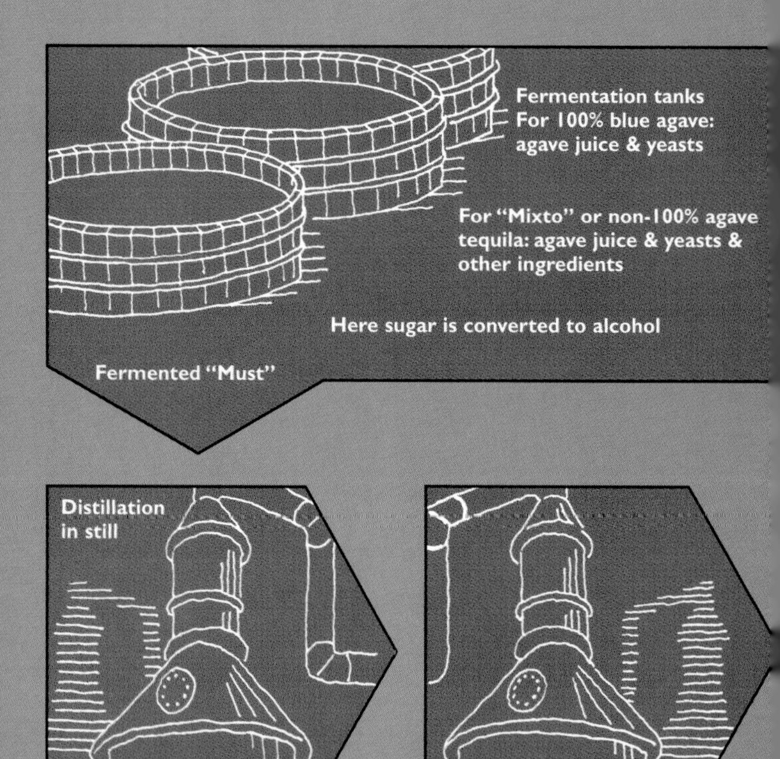

1st Distillation

2nd Distillation

Non-100% agave tequila is transported to central locations in Mexico and the United States, where it is bottled.

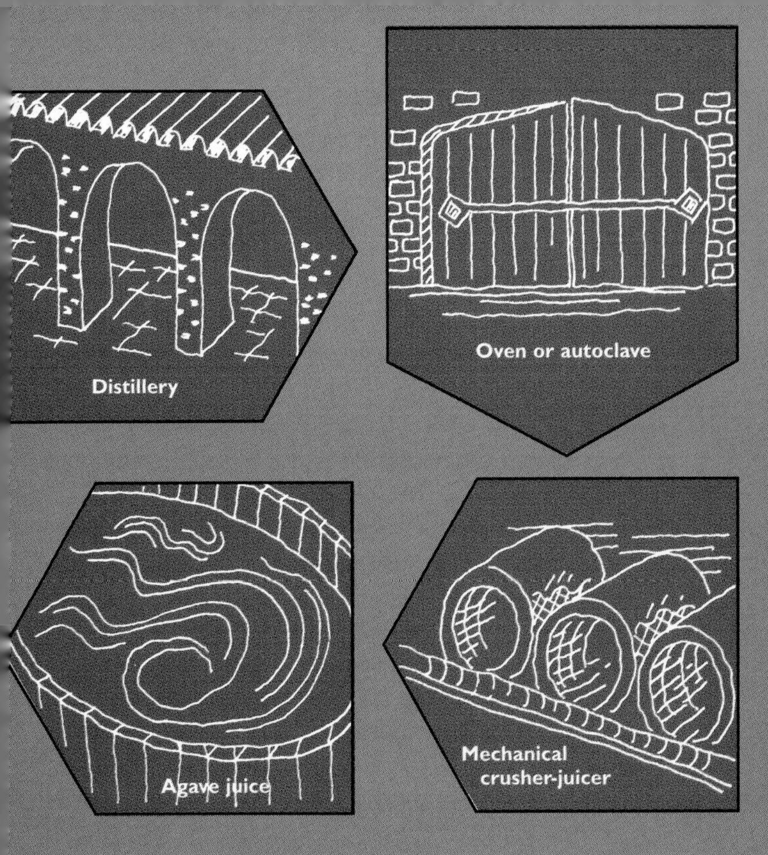

A. 100% Agave
Blanco
Bottled

Tequila

B. *Reposado*=Aged
minimum 2 months
C. *Añejo*= Aged
minimum 1 year

All 100% agave
tequila must be
bottled in
Mexico.

D. *"Mixto" Blanco*=
Unaged
E. "*Mixto* Gold" (with
added colors and flavors)

those tequilas that are called "100 percent agave" and those that are not 100 percent agave and may be referred to as *mixto*, or mixed tequilas. According to Mexican law, in order for a product to be called tequila at all, it must be made from at least 51 percent fermented juices from the blue agave plant. The agave content may actually be anywhere from 51 percent to 100 percent and the product can be called tequila. In order, however, for it to be called "100 percent agave tequila" the tequila must be just that. It must be made using exclusively fermented juices from the agave plant with no other sugars added. To avoid a negative connotation arising from the use of the prefix "non" as in "non-100 percent" or from the term "mixed," tequila producers prefer to refer to the two types as just "tequila" and "100 percent tequila." Mixed tequilas can contain anywhere from 51 percent to 99 percent juices from the blue agave. The remaining 1 percent to 49 percent will be other sugars added during the fermentation process, most likely cane sugar or molasses.

One hundred percent agave tequilas as a group are more generally referred to as premium tequilas. Especially in the case of several new small producers, the aged products appearing on the market use the term "super premium." The second product may variously be referred to as just plain "tequila," "*mixto* tequila," "mixed tequila," or "blended tequila." In fact, there is no official designation for these products, but suffice it to say that if a label does not clearly state that it *is* 100 percent blue agave, then it is for sure a blended product.

Another requirement of the Mexican government for the production of 100 percent agave tequilas is that they be aged and bottled within Mexico. In order to produce officially designated 100 percent agave tequila, each distillery must apply for permission, and receive approval from the *Dirección General de Normas*, which is the federal office controlling tequila production. The *Consejo Regulador del Tequila* (CRT), a regulatory council set up by the government and the tequila producers, then closely monitors the production of tequila through its various stages by constant inspections and testing. Limiting tequila production to the specially designated areas in Mexico enables these supervising agencies to have greater control over the processes and the final product's quality. All of the tequilas of the premium variety leave Mexico bottled and packed in

Pictured on the far right, tanker trucks called* pipas *transport large quantities of undiluted, high-alcohol tequila to bottling locations near central distribution points in Mexico and the United States.

cases ready for shipment to their final destinations.

The bottling requirements for mixed tequilas are different from those for 100 percent agave tequilas. Only a few mixed tequilas are bottled in Mexico at that point of production alongside the 100 percent agave tequilas. This occurs often in the same facilities where the tequila is kept separate and is bottled during separate production runs. The rest is shipped out of Mexico in tanker trucks, called *pipas,* at a higher alcohol level, as "bulk tequila." This bulk product will likely end up in bottling plants in the United States near key distribution points, where the product is diluted down to the appropriate alcohol content and bottled for domestic consumption.

Tequila connoisseurs will no doubt argue the merits of only 100 percent agave tequilas and will frown upon consumption of anything else. Nevertheless, many people are consuming the "other stuff," and in reality there are good tequilas among both types. For a true "tequila experience," a 100 percent or "real tequila" will most effectively convey the blue agave's flavor and its true character. Typically, these tequilas will have a superior bouquet, body and finish. It would be fair to state that the addition of other sugars, which may to some extent mask the agave flavor, gives the producer of blended tequilas greater latitude in its agave selection. Most importantly, different tequilas are better suited for drinking in different ways. Many of those possibilities are explored later in this book.

The increasing awareness about tequila has enriched the public's appreciation of the qualities of 100 percent agave tequilas, which make up the growing premium or luxury segment of the market. Keeping things in perspective, it should be noted that still less than 5 percent of all tequila exported from Mexico is of the 100 percent agave variety.

TEQUILA CATEGORIZED BY AGE

Besides the level of agave content used to create a particular tequila, the other key determining factor in a tequila's quality and character—which are defined by its appearance, aroma, texture and taste—is its level and type of aging. Tequilas commonly consumed can range anywhere from one day to several years out of the still. The resulting range of characteristics of these products is dramatic, and they will be enjoyed and appreciated more after gaining an understanding of what their differences are and what causes those differences.

While in Mexico, I once asked a worker in a tequila factory how he described the taste of tequila. He answered my question with another question, asking me, "What does thyme taste like?"

"It tastes like thyme," was my response. Well, good tequila tastes like agave, and to know what that means you have to experience it. Once you begin to do so, the sensational flavor profile is unmistakable, and to my taste, far more interesting than other clear spirits. In the most general terms, when compared to other spirits, tequila has a more earthy, herbaceous taste. Depending on how long a particular tequila has been aged, the range of flavors will run from earthy—almost mushroom-like—to peppery and sometimes minty; and in the case of aged tequilas, of course, from vanilla to smokey.

Now that the basic differences for tequila production have been revealed, why is it that some clear tequilas, for example, sell for more than $25.00 a bottle while others sell for only $8.99? Here is where the important distinction between tequilas of the premium variety—which are made exclusively from 100 percent blue agave "juice"—and those that are mixed or of the 51 percent to 99 percent variety, comes into play. With non-100 percent agave or mixed tequilas, the flavor of the product will be affected by whatever other sugars were used in the "formulation" of that particular tequila. This is not meant to say that the mixed product is not good. However, when tasting tequila "straight," in order to appreciate the full essence of the agave, the 100 percent product allows the fullest appreciation. When tequila is going to be mixed in a drink, then any good quality tequila—of which many are available, mixed or not—may be a fine choice.

BLANCO TEQUILA (TYPE I) As described in the production procedures, after distillation, tequila in its pure form is *"blanco"* or "white" tequila. Sometimes this product is also referred to as "silver" or the Spanish *"plata."* There is absolutely no difference in the products carrying these different names, but rather in the case of this category, more than anything else, the different names reflect each producer's own marketing preferences.

Blanco tequila is tequila in its purest form. It is the product of the fermentation and then distillation processes without the effects of any barrel aging. Legally, a *blanco* tequila is any tequila that has not been aged for at least sixty days. In fact, almost all producers will store the tequila in only stainless steel tanks, as opposed to wooden barrels, for a short period of time before bottling them. There are very few exceptions, and in those cases, the tequila gets only very brief storage time in large wooden barrels. This process prevents little if any flavor from transferring from the holding tanks into the tequila, thus creating the purest product.

Blanco tequilas are clear in color and can be either 100 percent agave or they can be mixed tequilas. Since the distillation process removes all impurities from the product and by definition distills the alcohol out of the product, everything except for the alcohol is in effect removed, thereby rendering this clear product. Since the flavor of the natural tequila has not been altered by any contact with wood, *blanco* is the most pure expression of tequila and its true essence, that of the blue agave plant.

The aroma and taste of a *blanco* tequila is the most natural expression of the agave in the beverage form. Typically, *blanco* tequilas will have a floral, herbal and somewhat peppery quality that is balanced by the agave's natural sweetness. Generally, *blanco* tequilas have more "bite" than their wood-aged counterparts. This inherent characterisitic is very often termed dry, or even spicy or peppery. A good *blanco* tequila that is made from 100 percent agave can actually have a great deal of finesse as it displays its crystal clear flavors.

REPOSADO TEQUILA (TYPE III) *Reposado* translated literally means "rested." By law, a *reposado* tequila must be aged in wood for at least sixty days. They are almost always aged for less than one year. Sometimes this aging is done in very large wooden storage tanks, which can range in size anywhere from 10,000 to 30,000 liters, but in some cases, small oak barrels are used. While sixty days is the legally required minimum aging time, depending on a particular producer's style, the aging period of a *reposado* typically ranges anywhere from two months to nine months. I have even tasted *reposados* that were aged as long as thirteen months in wooden tanks, but that is unusual.

Aging affects the appearance, the nose or aroma, as well as the taste of the tequila. The initial wood contact changes the tequila from a clear liquid into a light straw or even medium golden color. The specific change depends on the type of wood used and the length of time aged. Technically speaking, the effect of the aging process is to oxidize the alcohols that change the chemical structure and make the product softer. This is different from *añejos,* where the main impact on the tequila from its aging is the transfer of the wood's characteristics to the tequila. In practical terms, in both nose and taste, the affect of the shorter aging period of the *reposado* is to begin to smooth the tequila or to "take the edge off." As a result, *reposados* taste less harsh and slightly mellower than *blancos.* However, because of the modest aging time, the effective flavor of the agave is still very pronounced. *Reposados* strike a nice balance between the essence or spirit of the blue agave, which is expressed so purely in a *blanco,* and the subtle influence that the oak imparts in barrel aging, which is usually characterized as vanilla and spice. The resulting effect of wood aging may be a pungent scent and a deeper, lingering flavor.

While very popular in Mexico, *reposados* were until very recently the least well-known category of the premium tequilas being exported. Again, with the growing

The bottling line at Tequila Herradura where both the El Jimador and Herradura brands are packaged for shipment around the world.

awareness of and appreciation for tequila in general, that interest has extended to *reposado* tequilas and their fine nuances. There are now many excellent examples of *reposados* on the market of both the 100 percent agave type and the mixed types. Just like *blancos*, *reposados* can be enjoyed sipped "straight" or in mixed drinks. One example of a very popular drink in Mexico is based on a mix of *reposado* tequila with grapefruit soda. The same flavor differences that are noticed when drinking a *reposado* straight will carry over into their use in mixed drinks.

AÑEJO TEQUILA (TYPE IV) *Añejo* translated literally means "aged." According to Mexican law, an *Añejo* tequila must be aged for a minimum of one year and it must be aged in government-sealed barrels that are no larger than 600 liters. While one year is the legally required minimum aging time, anywhere from one to three years, depending on the particular distillery, may be common. Typically, the barrels used are old 190-liter whisky barrels from Kentucky. As a result of the longer aging period, an *añejo* tequila has a color that is darker than a *reposado*, and a nose and flavor that is even smoother and mellower. At this point the tequila takes on more of the qualities of other aged spirits like bourbons and whiskies, as the wood barrel has a greater chance to affect the body and flavor profile of the drink.

The best *añejo* tequilas still maintain and respect a delicate balance between the essence of the agave and the flavor imparted by the oak aging. Of course the effect of the oak aging is greater and more evident in an *añejo* then in a *reposado*. The object in making a fine *añejo* is to combine the agave flavor with the soft texture, vanilla flavor and spiciness that the oak contributes. The result should be a luxurious beverage like other aged spirits, with a complex depth of aromas and flavors, which can be savored.

With their greater heft and richness, then, *añejo* tequilas may be enjoyed in the same manner as other aged spirits. They may be sipped from a brandy glass or even a *capita,* the familiar dessert wineglass. In a nice large brandy snifter the many nuances of both the agave essence as well as the qualities imparted from the oak barrel aging can be captured and best appreciated.

This is not to say that aged tequilas, whether they are *reposados* or *añejos,* should never be mixed into a margarita or other cocktail. In fact, aged tequilas create the ability to mix a whole other style of drink that many people prefer. At Mesa Grill we have long used the very popular Sauza Conmemorativo, an *añejo,* in our famous "Prickly Pear Margaritas" to give the cocktail a desired balance and smoothness. If quality ingredients are used, then chances are, the true qualities of any tequila will show through in a mixed drink, and you should let your own personal taste be your best guide. In the recipe section, we will have more fun exploring the many possibilities of using different tequilas in different drinks.

Like *blancos* and *reposados, añejo* tequilas may be either 100 percent agave or mixed. However, most of the *añejo* tequilas that are now available are of the 100 percent variety in order to take advantage of the strengthening market for the so-called "premium" tequilas.

For a short period of time in recent years, a few products reached the market with the designation *muy añejo* appearing on the label. The inference was of a product which was "very aged." A few tequila producers riding the wave of the explosion of the super-premium market decided to go a step further and market a product even more luxurious than the *añejo*. However, since the CRT had not declared this designation as official, all of the producers who were making products labeled *muy añejo* ceased labeling that way and returned to using simply the *añejo* designation. If the one-year minimum aging requirement has been exceeded, producers wish to explicitly note this on the label. If a product is aged for three years the label will state "aged for three years."

What has evolved more recently are several new products that do receive additional aging in different types of barrels, and the producers are giving them new proprietary names to differentiate them from their other *añejo* products. Examples are El Tesoro's "Paradiso," Herradura's "Seleccion Suprema" and Porfidio's "Barrique." Despite their slightly different production techniques and their significantly higher price tags, all of these new tequilas are still officially categorized as *añejos.* Since extensive aging does not benefit tequila the same way that it does many other distilled spirits, most of the difference in these super-deluxe products and their *añejo* brethren are in the type of wood used for the aging. Some of the premium producers have turned to France and are using the same barrels in which cognac or great wines are aged.

JOVEN ABOCADO (GOLD) TEQUILA (TYPE II) As already described, *blanco, reposado* and *añejo* tequilas may be either 100 percent agave or mixed tequilas, making a total of six different possible classifications or types of tequila discussed so far. The seventh and last official classification is the *joven abocado* or "gold" category of tequilas, which are almost always produced as mixed tequilas. Until the recent explosion of premium tequilas and the growing awareness of the qualities and attributes of the different types and grades of tequila, "gold tequilas," as they are commonly known, were the recognized standard that lined most liquor-store shelves and the back bars of most restaurants and bars.

Even today, approximately more than half of all tequila exported is *joven abocado* or gold tequila. Another third of exported tequila is *blanco,* most of which is mixed. For comparison sake, only three percent of exported tequila is *reposado* and another three percent is *añejo.*

Gold or *joven* (young) tequila is an unaged or *blanco* tequila to which additive colors and flavors are mixed in after the distillation process. Since distillation renders a clear product, in order for the tequila to take on a color by some means other than aging it in wood, it must be through the addition of coloring. This is most often done

Barrels for aging* añejo *tequilas are closed with seals that may only be broken by inspectors from the CRT.

by adding caramel. The object is to simulate the effects that aging would have had on the product, had the tequila actually spent time in a barrel.

The intent of the added caramel or other flavorings and colorings is really threefold. First, the additives take away some of the harshness typically associated with unaged tequila. While that perceived harshness may be a sign of purity and therefore a quality sought after by many tequila connoisseurs, more casual drinkers may prefer a more smooth flavor. Since "time is money" in any business, the addition of flavorings and colorings is far less expensive than the cost of actually aging the product in oak barrels, to say nothing of the extremely high cost of the barrels themselves. As a result, products that get this type of artificial or simulated aging can cost substantially less to produce than a genuine *reposado* or *añejo* product, which spends considerable time in expensive oak barrels. In turn these products will be less expensive. If the truth be known, gold tequila is mostly produced for the export market. Don't expect to hear many Mexicans using the word *oro* to order tequila.

Of course, the effect of actual tank or barrel aging yields a more authentic product with a more genuine flavor: the blue agave combined with the influence imparted by charred oak, for example. Such a delicate vanillin or oak flavor could never be created with the addition of artificial flavors and colors. Since many flavors become obscured by the other ingredients in mixed drinks, the consumption of gold tequila is best saved for that use where the true essence of the blue agave is not of paramount importance but rather is being combined with others as in the case of flavored margaritas.

A NOTE ON "BULK TEQUILA"

Economics dictate that much tequila is actually bottled outside of Mexico. By law, only 100 percent agave tequilas and all aged tequilas, both *reposado* and *añejo*, must be bottled in Mexico. That leaves all mixed *blanco* tequilas and all *joven abocado* tequilas that do not have to be bottled in Mexico. Those tequilas still happen to make up most of the volume of the export market. For these tequilas that are later diluted before bottling, distillation is completed to 55 percent alcohol or greater, rather than the precise 40 percent at which it can be immediately bottled or aged. By virtue of its sheer quantity, the production of this tequila, which is made and shipped "in bulk," is of great importance to the tequila industry as a whole.

If a tequila is going to be sold as a *blanco*, then it is shipped "as is." If it is going to be a *joven abocado*, then the caramel or other artificial coloring and flavoring are added in Mexico, again in accordance with regulations

Montezuma and El Toro, both from Barton Imports, are examples of two of the better quality "bulk tequila" brands. Available as white or "gold," they are marketed with a price-value strategy and are typically used for making "house" Margaritas in better bars and restaurants.

The old-fashioned way: At Tequila Tapatío's bottling facility located in the town of Arandas, a worker affixes a bottle seal by hand.

that attempt to assert a certain level of quality control on the production process. Either clear or colored, then, the tequila is loaded into big stainless steel trucks resembling oil tankers and is "bulk shipped" to the United States where it is bottled. Before bottling, the tequila will be diluted with distilled water down to the legal level of 40 percent alcohol by volume or 80 proof level. Great economics are achieved by shipping at the higher proof so that the eventual yield of bottled product is increased by the amount of water that is added in dilution.

In the case of most of the largest tequila producers, most of their business is actually done in bulk tequila, which they in fact bottle with their own labels under their own brand names. Many other large distilleries produce tequila in bulk for various different brands, which may appear in Mexico as well as on the export market.

MAKING SENSE OF THE TEQUILA LABEL

Whether you fancy yourself a wine connoisseur, a single-malt scotch expert or a beer aficionado, do not be ashamed if you approach a bottle of tequila and its label with some degree of confusion.

To understand what you are drinking and where it came from, you have to be able to differentiate between distilleries, producers, bottlers, importers, brand names or product names of tequilas, and finally, classifications with respect to a particular bottle's content and age designation.

DISTILLERIES There are presently approximately fifty distilleries or tequila factories in the approved areas that are making tequila. Each distillery is registered with and approved by the Mexican government, but because of the growth of the market the approximation reflects the ever expanding production base.

In most cases, the name of the tequila producing company is different from the given name of a particular distillery that supplies it with tequila. While the tequila company is often named after the people who own it now, or who may have owned it in the past, as in the case of Tequila Cuervo or Tequila Sauza, the distilleries or factories sometimes named on the bottles often have some greater historical significance. For example, the principal distillery where Sauza products are made is called "La Perseverancia." Each distillery is assigned a NOM number by the governing agency so that the origin of every bottle can be traced.

In the example shown [see page 60, label (1)], the name of the actual distillery is in small letters where it says "fabricated and bottled by: Jesús Partida Melendrez." This distillery where the product was made will correspond to the NOM number assigned to that distillery. In this case, the number 1258 corresponds to the Melendrez distillery [see page 60, label (2)]. As explained above, the distillery or plant has an additional given name, which in this case is "La Magdalena Factory Las Esmeralda Ranch." That name probably has some historical meaning to the facility's owners.

PRODUCERS/TEQUILA COMPANIES Usually the most prominent name on the label or bottle will be that of the "brand" of tequila contained inside. Sometimes this is simply the same as the name of the producer, but where a producer makes several products, the producer's name may only make up part of the name. In the example of Sauza, there are products such as "Sauza Conmemorativo" or "Sauza Hornitos" where Sauza is the name of the producer, and Conmemorativo and Hornitos are the names of separate and very distinct products.

In the case of the label shown, the name "Tres Mujeres" indicates the name of the tequila company as well as the particular brand of tequila contained in the bottle [see page 60, label (3)]. However, it does not necessarily tell us exactly where the tequila was made. Any tequila made in an approved facility could be called "Tres Mujeres" as long as the name and proper NOM is indicated on the bottle. In another example, the name "El Tesoro de Don Felipe" refers only to the brand name of the product. It does not refer to the actual name of the tequila producing company, which is Tequila Tapatío, or to the facility where it is made, called La Alteña. There, tequila is made under *another* brand name which is in fact named after the facility. That tequila brand does reflect the name of the company owning it and is called simply "Tapatío."

IMPORTERS In the modern world of spirits' distribution, the largest brands are usually owned by even larger international corporations. With tequila, that is the case with many of the large brands. They are in fact small divisions of larger companies. which import that product or brand into big markets, most notably the United States, along with other brands of other spirits or other products. In the case of smaller brands, there is usually a corresponding, more specialized importing company that imports and represents that brand for sale along with other small brands of other products. The actual web of brand ownership, product importation and distribution can become even more complicated. But suffice it to say that somewhere on the label will be the name of a party responsible for that product's eventual arrival on your local liquor-store shelf or restaurant back bar [see page 60, label (4)].

The point of this elaboration is to explain that with so many new tequilas coming to the market, if you should happen to find a new or rare one that you especially like, there is somebody to get in touch with if you are having trouble finding more. In the label shown, the importer is Trans Comercio, Inc.

TEQUILA TYPE Further designation on the tequila label refers to the type of tequila in the bottle. If a tequila is 100 percent agave, then the label will surely say so [see page 60, label (5)]. If the tequila is a *mixto* or non-100 percent agave tequila, there will be no clear mention of that fact. However, any reference to 100 percent will be conspicuous by its absence. With respect to a tequila's age, if a tequila is a *reposado* or an *añejo* than the label will clearly say so [see page 60, label (6)]. Many *blanco*, or white tequilas, particularly those of the premium variety, will clearly state their age (or lack thereof) as well. Gold, or *abocado*, tequilas may not clearly state what they are, but the absence of any other designation, in conjunction with their color—simply ask yourself how this tequila got to be this dark color with no age—will clearly identify them.

DRINKING AND TASTING TEQUILA

The increasing awareness of tequila's distinct categories and the virtues of its better quality products has changed the way people are drinking tequila. When the choices were limited to predominately mass produced, inexpensive, less authentic products, then the image of "slamming" shots might have been a fair portrayal of typical tequila consumption. Now tequila has become a premium spirit worthy of both savoring and combining with other fine ingredients to make premium cocktails.

While I have nothing against shot glasses, it just so happens that in my tasting the many different tequilas now available, a shot glass will not provide you with the greatest opportunity to taste the nuances that give the various tequilas their character. To use an analogy, one would not recommend knocking back a shooter of single-malt scotch, and so similarly, good tequila should be savored and not shot. Knock back a shot of tequila in a single gulp, and you will miss its subtleties.

It is not the shape or size of a shot glass that is the problem, rather it is the tendency to shoot, rather than sip and savor the tequila when "shooting" it. In Mexico, tequila is often served straight, in shot glasses that are taller and more tapered than the standard shot glass. The shot is often accompanied by a second glass containing a *sangrita* chaser. *Sangrita* is a spicy mix of equal parts orange juice and tomato juice; lemon or lime juice; salt and a hot chile pepper "sauce."

Whatever type of glass you choose to use when drinking your tequila, always use the same method—let the tequila hit the tip of your tongue, wash over your taste buds and fire your palate. With a basic understanding of the general tequila characteristics, the following are some suggestions about how to approach this "mystical" drink:

Blanco tequilas (also known as *plata* or silver) will give the greatest impression of fruit (the agave character) and by virtue of their purity, are the most biting tequilas. If you must use a shot glass, then do so with *blanco* tequilas—the flavor is hard to miss. Because of the intensity of bright flavor in a *blanco,* it might better be enjoyed or appreciated in a rocks glass, either straight or with ice. Another thing I often do with *blancos* is keep them refrigerated and serve them chilled. The temperature helps to smooth out some of the harshness without compromising all of the great flavor.

At the opposite end of the spectrum are the *añejo* tequilas. These have a much greater complexity then the *blancos,* and should not go anywhere near a shot glass. Generally, *añejos* are consumed like other luxury "brown" spirits. Some people prefer them on the rocks as their cocktail of choice in the same way that they might consume a single-malt scotch or small batch bourbon; others feel that the complexity of the *añejo* is more befitting of a snifter in place of a cognac after dinner. A snifter is ideal for capturing the unique nose of both the *añejos,* and the even richer and more flavorful "super *añejos.*"

Somewhere in the middle of the spectrum are the *reposados,* which I favor for their flexibility. *Reposados* are characterized by their nice balance between a *blanco's* biting fresh agave fruit flavor, and the added smoothness in flavor and texture of an *añejo.* A *reposado* may be enjoyed in any of the same ways as a *blanco* or an *añejo.* Try a *reposado* straight, on the rocks or in a mixed drink.

Speaking of mixed drinks, what about Margaritas? After all, the great popularity of the Margarita cocktail is in no small part responsible for tequila's growing celebrity. My personal preference in a Margarita is to use a *blanco* or a *reposado,* because the bright flavors of these tequilas stand up against, and best complement, the fruity freshness of the other ingredients. Use an *añejo* for a smoother and more complex Margarita. (See the section beginning on page 145 for more discussion on the Margarita).

El Tesoro
de Don Felipe
SILVER
AÑEJO
PATRON
SILVER
PATRON
TEQUILA REPOSADO
40% Alc. Vol.
NOM 1258
Tres™
Mujeres
MR
LA MAGDALENA FACTORY LA ESMERALDA RANCH
TEQUILA
BLANCO
AÑEJO
PREMIUM
Premium
JOSE
CUERVO
Tequila
Cuervo Especial
Tequila
Cuervo
1800

TEQUILA PRODUCERS

"Tequila Producers" looks at today's tequila market on a brand-by-brand basis. The tequila brands included were chosen to highlight one or all of the following elements: colorful production methods, innovative processing techniques, and intriguing and romantic stories. Each of these brands lends a unique angle and perspective to an industry that is rich with history but also continues to evolve.

On publication of this book there are approximately fifty distilleries producing tequila in Mexico. Many of the distillery names would be unfamiliar to even the experienced tequila drinker, because the companies that own the distilleries are primarily in the business of producing tequila for sale as "bulk tequila" under various brand names; other companies produce tequila only for the domestic (Mexican) market. At the same time, there are several tequila brands, premium brands included, which actually contract with different distilleries to make their product for them, and those brands' producers may actually change from time to time as well. Those distillery names would be unfamiliar, too. There are, of course, still a number of companies who own distilleries and use their own names to brand their products.

With new brands entering the market all the time, the listing here in no way purports to be complete. Rather, I have selected a handful of the more popular brands that are either currently receiving or will soon be receiving national distribution in the United States. Obviously there are a plethora of additional brands which are available only in Mexico, as well as a number of additional brands that are available for the most part only in selected markets. I have mostly favored the inclusion of premium tequila producers rather than an exhaustive list of all the bulk tequila producers—who are really far less distinct from one another.

Listed in alphabetical order, each brand description includes some remarks about the brand's histories as well as information about the company's production processes and notes on their specific products. In putting each specific product in one of the regulated age categories, I have noted the amount of aging provided for that specific product according to the company. Accurate information about aging is sometimes sketchy, which is one of the reasons that there is no substitute for tasting the tequila itself. *Blanco* tequilas are typically unaged but legally can be aged less than sixty days; *reposados* are aged a minimum of sixty days; and *añejos* a minimum of one year. Joven abocado tequilas are typically unaged, but may be artificially colored. To contribute some flavor to the descriptions, I have incorporated notes and reactions from the tequila tastings conducted with the Mesa Grill tasting panel in early 1998.

CENTINELA

COMPANY AND HISTORY

Tequila Centinela, located in the town of Arandas, is one of the producers making exclusively 100 percent blue agave tequilas which are located in the Highlands of Jalisco in the area known as *Los Altos.* Since 1894, Centinela has been produced at a small- to medium-sized, family-owned distillery believed to be one of Mexico's oldest, having received its license to produce tequila in 1904.

When I visited the distillery, it was undergoing renovations and in the process of expanding its production capacity in order to meet the growing demand for tequila. Like many of the tequila producers which produce different brands for the domestic and export markets, Tequila Centinela makes another 100 percent agave tequila called El Cabrito, which is distributed almost exclusively in Mexico, with only limited distribution in border states such as Texas and California. The Centinela brand is sold both domestically in Mexico and exported to the United States. While the two brands produced at Tequila Centinela are quite similar, there are some processing differences that allow the much larger El Cabrito brand to be produced on a shorter production schedule. While both products are of excellent quality, Centinela remains the company's superior brand.

After appearing in the United States on a very limited basis in the 1930s and 1940s, Centinela was reintroduced slowly in 1993 and in earnest in 1995. It has since built a very loyal following among tequila connoisseurs. Master Distiller Jaime Antonio Gonzalez Torres, who has been making tequila since he was twelve years old, has supervised Centinela production for more than thirty years. He learned his trade from his uncle Salvador who distilled Centinela from the 1930s to the early 1960s. His antecedents began distilling tequila more than 100 years ago. Rooted in that history, the company prides itself on its use of many traditional techniques to produce a genuine product of consistently exceptional quality.

At Tequila Centinela, old and new technologies are married to create an authentic-tasting product. Agave is baked in an old-fashioned* hornos *before its juices are extracted using a modern agave milling machine.

PRODUCTION AND PRODUCTS

While Centinela owns some of its own agave fields, they also buy a portion of their agave on the open market, using only blue agave grown in the rich, red soils of the Highlands of Jalisco. Centinela's guiding maxim is *"Unico por su pureza,"* which means "Unique for its purity." Many things are still done the old-fashioned way, and technology is utilized only where it is believed to be beneficial to enhancing the product. Only stone ovens, or *hornos,* are used for baking the agave. After baking, the agave is milled or shredded, and all fermentation is done exclusively with natural yeasts. No chemicals are added to speed up the fermentation process, which typically takes eight to ten days in 10,000-liter stainless steel fermentation tanks. After double distillation in stainless steel pot stills, the four different Centinela products are bottled or aged. Each has its own special character and its own loyal following.

CENTINELA BLANCO

Type: 100 percent blue agave
Aging: **Blanco;** no aging

An unaged, very pure tequila. For the true *Los Altos*-type agave flavor, the Centinela Blanco is an excellent barometer. Tasters found this tequila to be very smooth with a mild agave flavor, making it very accessible, especially for the less-experienced tequila drinker.

CENTINELA REPOSADO

Type: 100 percent blue agave
Aging: **Reposado;** aged six to nine months

Centinela ages their *reposado* for between six to nine months, well in excess of the two months required for the category. Resting in small 200-liter white American oak barrels, which were previously used for aging whisky in Kentucky, the Centinela *reposado* acquires a light touch of oak flavoring, and the tequila is generally characterized as being "delicate." I would drink this tequila straight or even mix it in a premium Margarita.

CENTINELA AÑEJO

Type: 100 percent blue agave
Aging: **Añejo;** aged 12 to 18 months

As with their *reposado,* Centinela gives the *añejo* longer than the requisite aging to clearly define its category characteristics. Barrels that were previously used to age the *reposado* are later used to age the *añejo,* if the barrels are deemed to meet Centinela's standards. Because the barrels are being used to age tequila for the second time for the *añejo,* the resulting product is actually less tannic and smoother with a longer finish. In this case, the barrel aging serves to smooth out the tequila and to actually impart a woody flavor. Tasters responded favorably to the sweet, soft, lovely nose of this *añejo* tequila, which maintains a superb agave flavor. Centinela Añejo is an excellent sipping tequila.

CENTINELA AÑEJO "TRES AÑOS"

Type: 100 percent blue agave
Aging: **Añejo;** aged three years

Centinela's "super" product is the "Tres Años." This is actually an *añejo* which, during its aging process, is specially chosen as being best suited in character to handle the additional aging. This product is considerably less expensive than many of the other entrants in the "very aged" or "super" category. This tequila, with its additional aging and the oak's vanilla-flavored influence, gets farther and farther away from the spirit of the agave, but it has proven to be very popular with drinkers seeking a warm, complex spirit with plenty of fruity, agave flavor for drinking from a snifter.

From left to right: *Centinela Blanco, Centinela Reposado, Centinela Añejo and Centinela Tres Años.*

CHINACO

COMPANY AND HISTORY

From the distillery La Gonzaleña, and with a very rich history, the Chinaco brand is notable for several reasons. But before telling the fascinating story behind the brand, since this is a "guide to tequila," it would be remiss to not state at the outset that Chinaco's tequilas are amongst the best and most interesting of those being distributed in the United States today.

The Chinaco brand, and La Gonzaleña's existence as one of only two tequila distilleries located and operating outside of the state of Jalisco, has colorful roots in Mexican history. The Chinacos were a group of wealthy landowners who joined together with their workers in the fight for Mexico during the War of Reform in the 1850s and the French Intervention in 1863. These men, known for their bravery as well as for their elegance, were led by a man named Manuel Gonzalez. After the wars, with his military rank of general, Gonzalez returned to his birthplace, the northern state of Tamaulipas. He then began buying extensive tracts of land from Tamaulipas to Mexico City, which enabled him to establish valuable contacts at Mexico's Department of Agriculture.

Gonzalez was elected president of Mexico in 1880 and served for four years during which time he had many great accomplishments. Gonzalez brought electricity to Mexico City, and under his direction, the first Mexican-owned bank, Banco Nacional de Mexico, was founded. He also earned the name "Father of the Railroad," since the size of the national railway system doubled during his time in office. Gonzalez was later elected governor of Guanajuanto, and after his death, he was hailed by his country as a great Mexican hero.

General Manuel Gonzalez had a grandson named Guillermo Gonzalez. Guillermo began as a lawyer in Mexico City, later farmed land in Tamaulipas that he inherited from his great-grandfather, and ultimately was named the head of the Department of Agriculture. In 1965 Guillermo, assuming his new governmental post, went to Tamaulipas to survey the devastation caused by Hurricane Beulah. He found that some wild agave plants were the only vegetation remaining after the hurricane's devastation. Excited by what he saw, Guillermo contacted one of the large tequila producers in Jalisco—who had designed other distilleries—to help him convert the premises into a tequila distillery.

When Guillermo attempted to get his production permits approved, he ran into great resistance from the tequila establishment in Jalisco. Several large producers led the resistance, citing the laws that then existed governing the production of tequila and limiting it to Jalisco and the neighboring states. In the Chinaco tradition of his great-grandfather, Guillermo went to battle. In 1976 he gained the sympathy of the new Mexican president, Lopez Portillo, who agreed that the legal area for tequila production should be expanded to include the eleven municipalities of Tamaulipas. The additional area was officially recorded into the *Norma* in late 1977.

Tequilera La Gonzaleña soon began producing small amounts of tequila, which Guillermo named Chinaco after his great-grandfather's and his own battles, particularly his own triumph over the large tequila producers who saw him as unwanted competition. With limited production, Chinaco tequila was never marketed in Mexico, and its distribution was limited, too. The tequila was only available to the upper class in private clubs. Finally, in 1983 Chinaco was first exported to the United States where it has established a cult following of sorts.

Guillermo's four sons bought the distillery in 1993 and they have upheld all of the standards set by their father in producing only the most exceptional tequilas. While previously only aged tequilas had been exported under the Chinaco brand, in 1994 the first shipment of *blanco* produced by the new generation was exported to the United States, and after several rounds of appearing and disappearing from the market place, there is now a full selection of truly exceptional Chinaco tequilas available.

PRODUCTION AND PRODUCTS

Any plant will yield a product reflective of where it is grown. Wine may be the most widely used example of this principle. Different flavor characteristics are ascribed to wines that are made from the same grape but are grown in varying "appellations." The agave plant follows the same theory as grapes—agave grown in Jalisco's Lowlands or Tequila valley contrasts in flavors with Highland or *Los Altos*-grown agave. Tamaulipas, north of the other two areas, is an agricultural region with rich, brown soil that produces tequila with a distinct flavor of its own. All of the Chinaco tequilas are characterized by their great intensity of flavor with a very rustic, earthy quality.

While I have never visited the La Gonzaleña distillery, I am told by Chinaco's United States importer , Robert Denton, about the facility's modest scale. The distillery is described as utilitarian, because it was built "on the fly" in the late 1970s, when there was no ready outlet for all of the existing agave that had been grown. There is nothing fancy about the Chinaco facility. With no local tequila-producing industry to rely on, all of the equipment was trucked in from Guadalajara. At La Gonzaleña, agave is cooked in a small autoclave rather than a *hornos,* and distillation takes place in just two copper stills. The stills used here are even smaller than the ones pictured in the Tequila Tapatío facility where El Tesoro is made (see pages 82-90). While mostly modern equipment is utilized, Chinaco is essentially manufactured using traditional processes with only natural ingredients.

The terms most commonly used by the tasters for Chinaco tequilas were "earthy," "smoky" and "authentic tasting." Robert Denton explains that at La Gonzaleña there are dirt floors that are kept wet to maintain the humidity levels of the plant. The result is an earthiness which emanates into the barrels from the ground, naturally affecting the flavor of the tequila. The porous composition of the barrels encourages the infusion of an earthy flavor.

Recently, in the fast-evolving race toward more inventive bottling, Chinaco changed its packaging from tall, thin bottles which were used for so long, to new hand-blown bottles.

CHINACO BLANCO

Type: 100 percent blue agave
Aging: **Blanco;** no aging

Chinaco's *blanco* tequila garnered very strong reactions in my tasting, as most found it to be remarkably

From left to right: *Chinaco Blanco, Chinaco Reposado, Chinaco Añejo and Chinaco Reposado in the company's new handblown bottle.*

complex for an unaged tequila. "Herbaceous," "citrusy" and "almost smoky" were typical comments describing this most interesting of *blanco* tequilas.

CHINACO REPOSADO

Type: 100 percent blue agave
Aging: **Reposado;** aged eight months

The *reposado* from Chinaco is aged in small barrels. Tasters remarked on the tequila's beautiful nose and depth of agave flavor, citing such flavors as "chile," "smoky" and "a little salty." This smooth tequila does an exemplary job of maintaining a still-citrusy, agave character, and melding it with the effects of oak-barrel aging.

CHINACO AÑEJO

Type: 100 percent blue agave
Aging: **Añejo;** aged three to four years

In step with the two previous Chinaco tequilas, the *añejo* was one of the more exceptional sipping tequilas that the panel tasted. Chinaco Añejo generated extensive comments on the "presence" of this very bold tequila with abundant agave flavor. The following are a sampling of the remarks: "smells great, much darker in color than others"; and "not as soft—just a little hot, caramel flavor." This is one tequila, so impressively complex, that I would make a point of suggesting that it *never* gets mixed, but for real appreciation, should be sipped straight or on the rocks. Luxury!

DON JULIO

COMPANY AND HISTORY

Don Julio is actually the name of a brand of tequila that is produced by a company called Tres Magueyes. The Tres Magueyes distillery is located in the town of Atotonlico el Alto, in the Highlands of Jalisco, just up the road from another auspicious distillery, Siete Legues where Patrón used to be manufactured. Tres Magueyes is a company of great renown in Mexico where they sell tequila under both the Tres Magueyes brand name as well as the Don Julio name. Only Don Julio is presently being exported to the United States. Under the Tres Magueyes name, the company bottles a *blanco* as well as a *reposado* but under the Don Julio brand name there is presently only a *blanco* and an *añejo* being made.

In 1942, when Don Julio Gonzalez was seventeen, he founded the company. Like so many people living near the town of Tequila, he has worked virtually his entire life in the tequila trade, and he is thought of as one of its foremost authorities. Today, Don Julio Gonzalez still oversees the production of his tequila with the assistance of the members of his family.

As the company tells it, Don Julio kept his special private reserve tequila for friends and family until ten years ago when Reserva de Don Julio (the technical full name of the brand) was launched in Mexico. Only very recently, in late 1997, has the product been released to the United States, beginning in a select few markets. Tres Magueyes' recent alliance with the Rémy-Martin company for marketing and distribution of their brand should give them a jump-start in their new market.

PRODUCTION AND PRODUCTS

Until recently, Tres Magueyes was one of the largest exporters of bulk tequilas that were marketed in the United States under various brand names. In recent years, however, the company has shifted its emphasis to the production of 100 percent agave tequilas, including the Don Julio brand. The company's hands-on approach and their commitment to the premium tequila market has maximized their potential for success in the United States.

Don Julio has introduced two products to the market, Don Julio Silver and Don Julio Añejo.

With complete control over its production, Tres Magueyes should have no trouble meeting increases in demand for its product. In addition to owning their own factory, Tres Magueyes controls over 1,700 acres of agave fields ensuring a steady supply of rich Los Altos agave for their use.

DON JULIO SILVER

Type: 100 percent blue agave
Aging: **Blanco;** no aging

The Don Julio Silver is bottled fresh out of distillation with no aging. It benefits from the same soil characteristics as other tequilas do that are made from the Los Altos agave with its higher sugar content. The result is a "sweet-ish" vanilla agave flavor, which is mildly herbaceous but spicy at the same time—almost as if it had seen some barrel aging. Tasters enjoyed this tequila with its very attractive nose, but they wished that it "finished" more forcefully to complement that strong first impression.

DON JULIO AÑEJO

Type: 100 percent blue agave
Aging: **Añejo;** aged at least one year

Beginning as a Don Julio Silver and then aged in small oak barrels, the Don Julio Añejo is actually the first Don Julio product released to market after many years of producing under the Tres Magueyes and other brand names. Without extended aging, this product maintains a great agave character, and the introduction of the wood influence is not overpowering, but instead adds character and finish. Tasters liked this tequila even more than the silver, remarking on its "mocha-like" or even "chocolatey" characteristics. Obviously, this tequila is best sipped straight or on the rocks.

EL CONQUISTADOR

COMPANY AND HISTORY

Among the tequila brands listed in this book, El Conquistador is one of the newer premium tequilas on the United States market as the march of new tequila brands to market continues apace. Seizing on the trend toward innovative packaging, this new brand has been introduced by yet another large liquor company in the United States—Heaven Hill Distilleries of Kentucky.

Named after Cortés and the Spanish Conquistadors, who brought the distilling process to Mexico, the El Conquistador brand is actually the product of a tequila manufacturing company called Agroindustria Guadalajara and has previously been available in Mexico. A slightly confusing scenario exists around this brand: the Heaven Hill company had already been distributing a low-priced, mixed tequila—also called El Conquistador—from a completely different distillery, which had nothing to do with Agroindustria Guadalajara, when the latter company created their new premium product and decided to call it El Conquistador as well. Rather than change the name of the product, Agroindustria Guadalajara made an arrangement with Heaven Hill for exportation to, and distribution in, the United States.

The factory of the Agroindustria Guadalajara company is located in the village of Capilla de Guadalupe in the Highlands area of Jalisco, west of the town of Arandas. This is an area where a number of other more established brands are from, and where the fertile red soil produces the biggest, ripest blue agave plants. Less than five years old, the distillery is fairly new and has previously manufactured a number of other 100 percent agave tequilas available mostly in the Mexican market.

PRODUCTION AND PRODUCTS

Clearly positioned for the high-end of the premium tequila market, the exclusively 100 percent agave tequilas from El Conquistador fit right into that market with their attractive, expensive-looking packaging. All three El

***The three types of El Conquistdor tequilas:* blanco, reposado *and* añejo.**

Conquistador tequilas are packaged in tall, elegant, handblown bottles, each of which is unique.

Since I did not visit the distillery where the tequila is being made, I am without any firsthand knowledge of their production methods, and so I have relied on the company for information. Samples of the product (as it was bottled) were also made available to me. Clearly, in addition to carefully designing its image, this company knows a thing or two about making tequila. It has created products of impressive character and quality.

EL CONQUISTADOR BLANCO

Type: 100 percent blue agave
Aging: **Blanco;** no aging

Packaged in a cobalt-blue, handblown glass bottle, the *blanco* from El Conquistador is one of the few according to the company that is placed in oak barrels for brief "resting" before bottling. While evidence of wood aging was not noticeable, tasters did note strong agave flavor in this tequila, likening the nose and flavor to "dried fruit," but some found its first impression to be high in alcohol.

EL CONQUISTADOR REPOSADO

Type: 100 percent blue agave
Aging: **Reposado;** aged seven months

El Conquistador's *reposado* is aged for a minimum of seven months, well in excess of the legal minimum of two months. The tequila benefits from its aging as it acquires hints of vanilla and a smoother finish than the *blanco,* while still maintaining the fruitiness described above. In this well-balanced *reposado,* the oak aging has effectively removed the *blanco's* edge and has created a nice balance with subtle citrus and vanilla flavors.

EL CONQUISTADOR AÑEJO

Type: 100 percent blue agave
Aging: **Añejo;** aged eighteen months

According to the company, El Conquistador Añejo is aged in French oak barrels, rather than the more typically used American oak. The intention is to give the tequila a more unique flavor. The tequila benefits with a rather rich, aromatic and herbaceous bouquet, with a spicy well-balanced agave flavor, and what tasters recognized as "a long, slightly sweet finish." This is definitely a snifter tequila.

EL JIMADOR (see Herradura, page 102)

EL TESORO DE DON FELIPE

COMPANY AND HISTORY

While many tequila companies like to call their products "handcrafted," often the use of that term is a considerable stretch. In the case, however, of El Tesoro, there is no stretching the truth. At La Alteña, the distillery where El Tesoro is made by a company called Tequila Tapatío, vitually everything is done by hand. Visiting Tequila Tapitío is like visiting a working tequila production museum where most of the traditional methods of tequila production are still utilized every day. There is no gimmickry involved, as the truly artisan tequilas made here are without doubt among the best tequilas made anywhere. The name El Tesoro fittingly translates as "the treasure."

Located in the Highlands of Jalisco in the area called *Los Altos,* La Alteña was built in the town of Arandas in 1937 by Felipe Camarena Hernandez. At that time, Hernandez founded the company and actually built the distillery. He later passed the company down to his son, Felipe J.C. Camarena Curiel, who is the present director of the company known as Tequila Tapatío.

Tequila Tapatío currently produces tequila under two different brand names: El Tesoro de Don Felipe, named after the company's founder; and Tapatío, which shares its name with the company itself. Slightly more than half of the current production is the Tapatío brand, which is distributed exclusively in Mexico. The El Tesoro brand, making up the rest of the production, is distributed in Mexico, in addition to being exported to a small number of markets of which the United States is by far the largest. The newest and fastest growing market for El Tesoro is Japan, where people seem to be willing to pay for the best of anything and everything. There is no reason to think that tequila would be an exception.

While Tequila Tapatío is in the midst of a significant expansion of their factory, or *fabrica,* Carlos Camarena, who is Felipe's son and is in charge of production, insists that the company will continue to make all of their El Tesoro brand tequila only employing the old methods and using the traditional equipment. The new facilities will be used primarily for production of the Tapatío brand and possibly a new brand, which has not yet been created. At the same time, many of the traditional techniques, including baking the blue agave in stone and brick ovens, and the use of a *tahona* will be preserved as five new ovens have already been added to the two that were in place. Although there are plans to build an additional *tahona*, a more modern milling machine has also been installed and will be used to extract the juice from the cooked agaves for the newer brands.

PRODUCTION AND PRODUCTS

All of the blue agave used to make El Tesoro is "estate-grown" in Tapatío's own fields and is harvested by their own *jimadores*. Close oversight of the estate ensures that only the best, most mature plants make it to the ovens. After the *piñas* are unloaded at the receiving patio, they are loaded into the brick ovens where they are baked for twenty-four hours by steam, and then allowed to cool or rest for another twenty-four hours. On a visit to the distillery, I actually went inside the ovens at Tapatío and saw the eight locations at the top and bottom of the oven where the steam is ejected.

The cooked *piñas* are unloaded from the ovens into wheelbarrows, which are then wheeled down a plank and pushed from the wheelbarrows into a cobblestoned pit where, to my knowledge, Tapatío is the only company still using a *tahona* to crush its agave. The *tahona* is a gigantic stone wheel weighing about 2,000 pounds, which is attached by a long rod to a pole standing in the center of the pit, acting like a pivot. In a small concession to modernization, a mule or ox is no longer used to pull the *tahona* in circles around the pit; instead a tractor is used. While the *tahona* is grinding the agave to extract the agave juices or "honey water," a man called a *tahonero* stands inside the cobblestoned pit with a pitchfork, moving the crushed fibers around to ensure that all of the agave gets evenly crushed for maximum juice extraction.

After the cooked agave has been crushed by the *tahona*, the workers load the juice and the crushed fibers into small wooden buckets and carry them on

The tahona at Tequila Tapatío is the only one that I know of that is still in use. A 2,000-pound stone wheel circles the pit, crushing the agave while the tahonero moves the crushed agave around.

A worker carries the juice of the crushed agave, to the fermentation tank area along with the agave fibers in small buckets on his head.

Nothing fancy about this place. Workers are waiting for the final tequila product from the second stage of distillation.

their heads to the fermentation tanks. This procedure, with all of its tradition and charm, was not seen at any other distillery, and I am told that it basically does not take place anywhere else but at Tequila Tapatío. One of the most significant differences between Tapatío and other producers' methodologies is that Tapatío is the only distillery (at least the only one that I found) where fermentation is done without first removing the fibers from the agave juice after the crushing or milling of the agave. The agave fibers are left in, not only during the fermentation process but through the first distillation as well. It is believed that the fibers give more agave flavor to the *mosto* and eventually to the finished product. It is interesting to note that fine cognacs are also produced using many of the same methods as Tequila Tapatío.

Tapatío ferments in small wooden vessels that hold 3,000 liters each, rather than in large stainless steel fermentation tanks that can often hold up to 65,000 liters. The most surprising part about Tapatío's method was that I actually witnessed a man standing inside a fermentation vat as the juice and fibers were dumped in by other men carrying the small buckets of juice and crushed agave from the *tahona* pit. It is this man, called a *corralero*, whose job it is to make sure that the fibers are completely separated and that the yeasts, which are added, are thoroughly mixed and distributed throughout the tank to homogenize the mixture. At Tapatío no

A worker removes samples of aging tequila from a barrel in the distillery cellar. On the end of the barrels, you can see the stamp from Jack Daniels, Kentucky Bourbon—that's where these used barrels come from.

chemicals are used to accelerate fermentation, and only the naturally occurring yeasts—which are maintained at the distillery—are utilized. At this point, the sugar and yeast content are measured. One vat of native yeast is kept in the factory, and more is added to each batch as needed. This spontaneous fermentation process, depending upon the weather, can take up to five days before the *mosto* is finally ready to go to distillation.

At Tapatío there are presently thiry-eight fermentation tanks able to hold 3,000 liters each. But due to the presence of the fibers in the tank, there are only 2,000 liters of agave juice in each tank.

During the fermentation process, the yeasts "eat" all of

the sugars, producing alcohol and carbon dioxide. At the completion of the fermentation process, with the beginning of the production of alcohol, the product will have an alcohol content of 5 percent by volume.

At Tapatío, distillation is done in two small copper pot stills. The first distillation produces the *ordinario*, which is 18 to 20 percent alcohol and is not yet tequila. The second distillation is monitored very closely to bring the product to 41 percent alcohol content for the *reposado* and the *añejo*. Thereafter, the aging process will naturally bring the alcohol back down to 40 percent alcohol or 80 proof. For the *blanco*, the product will be distilled to precisely 40 percent, since there will be no dilution—either naturally or from the addition of water before bottling.

Tapatío cellars all of its tequila in an underground facility beneath the distillery. This cellar was designed by Don Felipe Camarena to reflect the famed cellars of Cognac, France. The cellars are deep underground with stone walls and arched brick ceilings. An ideal aging environment exists with cool temperatures and consistent humidity levels. Different types of used bourbon barrels function to provide the various tequilas with their individual characteristics.

In addition to the four El Tesoro tequilas discussed below, Tapatío produces two tequilas under the Tapatío brand name that are available only in Mexico. The *Tapatío Reposado*, which is aged for three or four months, and the *Tapatío Añejo*, which is aged for one year in oak, are both as exceptional as the El Tesoro tequilas, although they are more moderate in all of their characteristics.

EL TESORO DE DON FELIPE SILVER

Type: 100 percent blue agave
Aging: **Blanco;** no aging

Unaged, El Tesoro Silver is bottled just as it exists, directly out of the pot still after the second distillation and after a brief period resting in a large stainless steel tank. The use of the term silver is merely a substitute for *blanco*, or white, and is a name often applied to unaged tequila. All of the steps that are taken to make a more pure, genuine agave product for El Tesoro are in evidence here as the agave flavor jumps out of the glass before a sip is even taken. It has the most floral, richest agave nose of any *blanco* tasted. This fresh, very clean tequila has great characteristics of herbaceousness, even mintiness, as well as a bit of natural spiciness. This agave-derived spiciness is very different from the flavor that tequilas (and other aged beverages for that matter) adopt from exposure to wood. The tasters were most impressed by this tequila, noting its "genuine" quality; one remarked of being "transported to Mexico" while tasting. Others commented on El Tesoro Silver's balance, complexity and smoothness—all of which were noted as exceptional in an unaged tequila.

EL TESORO DE DON FELIPE REPOSADO

Type: 100 percent blue agave
Aging: **Reposado;** aged for nine months

This tequila was actually added to the El Tesoro brand line only recently, after the demand could not be met for El Tesoro Añejo, which needs considerable aging time. A new appreciation has developed for the entire class of *reposados,* and this example has already made its mark on the premium category. El Tesoro Reposado is aged for nine months in new, white oak barrels. The oak provides the tequila with considerable natural structure and color, as well as acting as a natural oxidant, softening the tequila. The robust nose reveals subtle nuances of oak and vanilla, while the taste benefits from the floral spiciness of the silver tequila, which is what it begins as. Still, it is mellowed from the softness of the barrel aging. Most of our panel were tasting this product for the first time, and the response was universally high praise. While many *reposado* tequilas make an excellent choice for use in mixing in Margaritas, I am inclined to reserve this one, with all of its flavor, as a great sipping tequila.

EL TESORO DE DON FELIPE AÑEJO

Type: 100 percent blue agave
Aging: **Añejo;** aged two to three years

Just as exceptional as the *blanco* and the *reposado,* El

Tesoro Añejo truly stands out from the crowd. The tequila is aged for two to three years, with the determination of exact aging time made based on monitoring the character of the particular barrels used. The same bourbon barrels utilized for aging the *reposado* are used here. Up to three years of barrel aging produces a luxuriously rich product.

A significantly darker tequila than the El Tesoro Reposado, the *añejo* product also stands out for its finesse, and as with the Silver, it has a notably minty character as it achieves a most harmonious balance of oak-derived vanilla with the fruitiness of agave. Try sampling this tequila next to another *añejo* and notice how powerful the sweet *Los Altos* agave is, especially when combined with the oak's vanilla influence. All tasters remarked on the harmony between this *añejo's* aroma and taste. This is a most exceptional sipping tequila!

EL TESORO PARADISO

Type: 100 percent blue agave
Aging: **Añejo;** aged more than three years

This luxury item is brand-new and should be arriving on the market at about the same time as publication of

El Tesoro's complete product line consists of El Tesoro Silver, El Tesoro Reposado, El Tesoro Añejo and the brand-new El Tesoro Paradiso.

this book. Fortunately, I was able to sample the Paradiso while in Mexico, as final arrangements were being made for its bottling, labeling and exportation.

This tequila is an innovative development, even among the new category of what I term "Super Añejos." The producers of El Tesoro joined with the famous cognac blender, Alain Royer of A. de Fussigny, to blend together various *añejo* and silver tequilas. The blend is first aged in old bourbon barrels, and then in French oak casks shipped from France after serving as aging vessels for cognac. The result is a tequila at the extreme. Notwithstanding its extensive aging, there is still the tequila's unique herbaceousness on the nose to perfectly match a sweet-from-the-cognac-barrel flavor—all of which revolves around the agave character, which is so pronounced in all of the El Tesoro Tequilas.

GRAN CENTENARIO

COMPANY AND HISTORY

The Gran Centenario brand is owned by, but produced, marketed and distributed completely separately from, Jose Cuervo. In response to the increasing demand for premium tequilas, Cuervo created the separate Gran Centenario brand of tequilas—which are all 100 percent blue agave tequilas—and at the same time discontinued its Dos Reales brand, which was popular with many tequila drinkers.

Gran Centenario is made in the Los Altos region in the town of Zapotlanejo. The tequilas are crafted from selected agave plants at the Casa Cuervo distillery known as Los Camichines—which is a Mexican estate known for its rich, volcanic soil and natural irrigation.

Los Camichines was founded in 1857 by Lazaro Gallardo, inventor of what he called *seleccion suave*, a process whereby only the smoothest, best quality tequilas are blended together in oak before resting.

Previously popular in Mexico, Gran Centenario tequilas have been available in the United States in their present form only since 1996. Longtime tequila drinkers will recognize the Gran Centenario bottle as looking amazingly similar to its predecessor, Dos Reales—a brand that has been discontinued.

PRODUCTION AND PRODUCTS

Compared to its parent Jose Cuervo, Gran Centenario tequilas are produced in relatively small quantities. They share many of the characteristics of other boutique tequilas as they utilize many of the older methods of production. Fully mature blue agave is baked in stone ovens, the juice of the agave is naturally fermented and distillation is done in small batches. If anything, the more traditional packaging is a refreshing departure from the variously colored and shaped bottles of tequila that are suddenly flooding the market. The brand's marketing strategy seems to be to create a distinctly classy look in keeping with the product's quality and price.

GRAN CENTENARIO PLATA

Type: 100 percent blue agave
Aging: **Blanco;** aged briefly

One of the few *blanco* tequilas that is actually aged very briefly in wood before bottling. The tequila is aged in specially designed new white oak casks just long enough to "take the edge off" of the tequila. The wood has a mellowing effect, imparting a very subtle flavor and aroma, giving the tequila a slight yellowish tinge. Still, this is definitely a biting tequila.

GRAN CENTENARIO REPOSADO

Type: 100 percent blue agave
Aging: **Reposado;** aged at least six months

This *reposado* is aged for at least six months in small white oak barrels—the same type of barrel aging that is applied to the *añejo*. The American whisky-barrel flavor is evident, as the tequila takes on a slight butterscotch flavor and finish.

GRAN CENTENARIO AÑEJO

Type: 100 percent blue agave
Aging: **Añejo;** aged at least one year

Its newest product, Gran Centenario's *añejo* tequila is a well-made, nicely balanced tequila. Gran Centenario has chosen to put their latest tequila in a newly shaped bottle, causing it to stand out from the other products in the brand. Aged for between one year and a year and a half in small, charred, white oak barrels, this tequila takes on a mature, mellow, full-bodied character with a smoky, single-malt scotch-like quality typical of the better *añejo* tequilas.

AGAVERO

Type: 100 percent blue agave
Aging: Not applicable

This unique, 100 percent blue agave liqueur is one of the new tequila-based products that seems to be hitting the market in the wake of tequila's growing popularity. While Agavero may be on the sweet side for a hard-core tequila drinker, it does have good agave character and a good story to go along with it. Made from a recipe passed down from Lazaro Gallardo himself, it begins with reserve barrels of *añejo* and *reposado* tequilas,

From left to right: Gran Centenario Plata, Gran Centenario Reposado, Gran Centenario Añejo and Agavero.

which are aged separately: the *reposado* for nearly a year; the *añejo* for no less than two years. The tequilas are blended with Agavero's so-called secret ingredient, a unique "tea" brewed from the Damiana flower. The Damiana, a small plant indigenous to the mountains of Jalisco, was used in Gallardo's day as a natural tonic. (Rumor has it that it is an aphrodisiac.) To Agavero, it imparts an intriguingly rich, earthy taste.

HERRADURA

COMPANY AND HISTORY

Herradura is the largest and most well-known of the exclusively premium tequila producers making only 100 percent blue agave tequilas. Tequila Herradura was founded in 1870 by Feliciano Romo and has been owned by the Romo family since then. The story is told of Feliciano's searching for a site for the construction of a new distillery when the glint of metal caught his eye. The glint was from an old horseshoe—thus the name chosen for the tequila brand was *Herradura*, Spanish for horseshoe, the symbol of good luck.

The company was run for many years by its prominent proprietress, Gabriela de la Peña Rosales Romo, until her death in 1994. The company is now run by her two sons and two daughters. Guillermo Romo is the company's general director, and he maintains the highest profile as he is a progressive fixture on the tequila scene in Mexico.

Herradura was first exported to the United States in the 1940s after being discovered by Bing Crosby and Phil Harris who arranged for its exportation to the United States.

Herradura is located in the Amatitán Valley, midway between Guadalajara to the east and the town of Tequila about six miles to its west and north. The Herradura plant was built around the Romo family's classic Hacienda—where rooms surround the open courtyard and garden. Located adjacent to the home itself is the requisite chapel (every hacienda must have one), where the ashes of Gabriela remain. It is not unusual to see any of the company's 900 employees regularly visiting the chapel for daily prayer or to pay their respects to the matriarch. There is a lot of history at Herradura, and the workers there all seem to be very loyal and proud. While visiting the facility, I met the family's cook, Doña Paula, who has been preparing meals for the Romo family for forty-six years! The facility is immaculate, with workers everywhere carefully maintaining the grounds. Tall fences and gated walls armed with security guards surround the entire compound. This level of protection is necessary due to the high profile of the Romo family. The family members, known to be outspoken on many issues in the tequila industry, often oppose the larger producers regarding production and economic matters.

The Hacienda itself takes its name, "Hacienda San José del Refugio," from the Guera Christea of 1922–1926 when the Mexican Government was at war with the legions of poor Catholic people, especially the Mexican *campesinos* or farmers. During the war many Catholics found refuge at the Hacienda, where they were allowed to hide out.

The grandeur of the Hacienda is indicative of the rich and complex history of the Romo family and is an emblem of the family's aristocratic social status. A private library, which sits inside the compound adjacent to the Hacienda, is actually the second largest private library in the state of Jalisco with over 23,500 books on its shelves. Also on display there is an impressive array of riding saddles, many of which were formerly the property of the former Mexican president, Benito Juarcz.

The growing popularity of the Herradura brand has made it necessary for the company to conitnually expand its production facilities. A third facility went on line in 1993, and construction is nearing completion on a fourth facility, which will increase production capacity by 80 percent. The first and oldest facility has been converted into a tequila museum with all of its old-fashioned equipment and apparatus on display. The fourth facility will soon join the second and third ones as fully operating plants.

The Herradura facility and production process are modeled after a small European "estate." The company grows all of its own agave and ferments, distills, ages and bottles all of its finished tequilas on the premises. Herradura's mark as a manufacturer of a tequila that is an all-natural spirit is an accurate portrayal. Not only is all of its production done without the use of any supplements, such as additional sugars, colors or flavorings, but only natural yeasts are used in the fermentation process.

Herradura enjoys the distinction of being one of the

very few true "estate grown" producers of tequila. This means that Herradura actually owns all of its agave plants and relies on its more than 500 *jimadores* to exercise full control over the harvest. This ensures that the best decisions are made with respect to harvesting at the precise moment in the agave plant's maturity, which in turn maximizes the concentration of flavors in the tequila that is produced. To my knowledge, Herradura is the only spirit authorized by the BATF to display the term "estate bottled" on its label.

The chapel at Herradura's San José del Refugio where company employees pray and pay their respects to the ashes of the family martriarch, Gabriela.

On weekends, members of the Romo family gather at the hacienda where the family cook has been working for the family and preparing meals for nearly fifty years.

PRODUCTION AND PRODUCTS

Gabriela's enduring attitudes toward agave culture have been described as "neoclassical," meaning that as the tequila-making craft evolved, she maintained for the company a traditional yet progressive posture toward the changes taking place. Guillermo has continued along the same philosophical lines as he continues the family tradition of making high quality, unadulterated tequila.

Even with its incredible growth, and despite Herradura's incorporating modern technology into the tequila production process, the most important steps continue to be done the traditional way, maintaining the importance of the product's authenticity.

Herradura still bakes all of its agave in its ten brick ovens, and while fermentation takes place in large stainless steel tanks, Herradura introduces no chemical yeasts. In fact, in several areas around the distillery, there are groups of Lima trees, which help to preserve the yeasts in the air that are captured and used in the fermentation process. After distillation, all of the aging of Herradura tequilas takes place in *bodegas* located within the confines of the estate.

Everything related to tequila production is done within the confines of the "estate" including the maintenance and repair of oak barrels used for aging.

HERRADURA SILVER

Type: 100 percent blue agave
Aging: **Blanco;** aged forty days

Most white or silver tequilas are unaged. However, Herradura Silver is a rare exception as it rests in wood for a brief forty days before it is bottled. The color of the tequila takes on a very slight yellowish tinge from its brief foray in the barrel, but there is no doubt that this is still a characteristically clear tequila. The aging is just enough to take some of the edge off and smooth out the tequila. Tasters responded to this tequila's pleasant nose, an attribute furthered by the time spent in barrels, which gives it a slight wood-spiciness on the palate, rather than wood-smokiness. A "citrusy" quality is well complemented by the same wood and spicy essence—the result of the very brief wood aging as well. This tequila's unique quali-

All of the agave used to make Herradura is farmed by the company's own jimadores. Because their farmers are not paid according to the weight of their harvest, they trim the leaves, or pincas, off as far down as possible, eliminating any potential bitterness.

ties make it incredibly flexible. Sipped straight or mixed in a margarita with the best ingredients, the tequila's best qualities shine through.

In Mexico only, Herradura bottles a different product (not listed and described separately here) called Herradura Blanco. While *blanco* and silver are generally interchangeable terms for unaged or clear tequilas, in the case of Herradura, they refer to two distinctly different products. Herradura's Mexican "Blanco" in this case is not the same as the United States "Silver," as the Mexican Blanco gets no aging and the tequila is bottled at 92 proof, considerably higher than the United States standard of 80 proof. Evidently, the company believes that Mexican tequila drinkers can better handle a harsher, more potent tequila than their northern neighbors. Further complicating matters is yet another product made exclusively for the Mexican market called Herradura Blanco Suave, which is an 80-proof product that is essentially the same as the United States Herradura Silver.

In any case, Herradura Silver, as it is sold in the United States, remains one of my favorite tequilas for sipping straight or on the rocks, and with its added body and flavor makes an exceptionally flavorful premium Margarita.

HERRADURA GOLD (REPOSADO)

Type: 100 percent blue agave
Aging: **Reposado;** aged for over one year

Herradura has created a slightly confusing situation by labeling their *reposado* tequila with the name "Gold." By now, the use of the term "Gold" is usually given to *joven abocado* tequilas, which are typically bulk-type, mixed, only 51 percent agave tequilas to which caramel coloring has been added to simulate the aging process. In the case, however, of Herradura, nothing is simulated or artificial. In the Herradura style, this *reposado* is aged far longer than the required two months. According to Herradura, this particular tequila is aged for over thirteen months, which could actually qualify it for categorization as an *añejo*. I would not be surprised to see Herradura removing the "Gold" label some time soon as consumers become increasingly aware of the distinctions between the various product categories.

Whatever Herraudra chooses to call its *reposado* tequila in the future, it remains a very distinctive product, with plenty of agave character and Herradura's trademark wood-infused flavor. This rich and flavorful *reposado* is complete from beginning to end—one taster called it a "*reposado* to the extreme." Tasters noted the wood character but still felt it was subtle enough to be considered soft and especially pleasing. The wood did not get in the way of the great agave flavor and what some tasters described as flavors of "almond."

Herradura, which is the largest producer of exclusively 100 percent blue agave tequilas, bottles their various products for shipment all over the world.

HERRADURA AÑEJO

Type: 100 percent blue agave
Aging: **Añejo;** aged two to three years

Herradura as a rule ages its products well beyond the prescribed minimum periods. Their *añejo* is no exception as it is aged more than twice the required one year. The effect of the extended aging is apparent in this full-bodied, rich tequila without being overpowering. Herradura Añejo impressed most tasters as an extremely appealing deeper-flavored product, although some tasters preferred the more balanced *reposado.* While the more flexible *reposado* could be drunk straight, on the rocks or in a mixed drink, this *añejo* would be much better sipped from a snifter.

HERRADURA SELECCIÓN SUPREMA

Type: 100 percent blue agave
Aging: **Añejo;** aged two to three years

Until the release of Porfidio's "Barrique," the most expensive tequila available with a retail price of nearly three hundred dollars a bottle, Selección Suprema is Herradura's version of "The Ultimate Tequila." With three to four years of aging in white oak, this tequila is purely a luxury item for sipping only. Four or more years of aging give this tequila a much greater resem-

Herradura at present is exporting to the United States its Herradura Silver, Herradura Gold, Herradura Añejo and Herradura Selección Suprema.

Herradura's answer to an XO Cognac, its Selección Suprema is one of the most expensive tequilas now available.

blance to a rich single-malt scotch than to the tequilas one is accustomed to tasting. In contrast to Herradura's signature square bottles, or the round ones used in Mexico, the Selección Suprema comes packaged in a heavy, molded glass decanter-like bottle with a glass stopper and the labeling embossed in gold. The bottles are then packed in boxes handcrafted from the bark of the amate tree, for which Amatitán, the hometown of Herradura, is named.

Like other Herradura products described here, Selección Suprema obviously shares the use of wood as a defining characteristic—in this case, to an extreme. With an almost sweet-flavor profile, this tequila, with its deep, dark color and rich, almost buttery flavor, is a wonderful complement to—or for my taste, substitute for—dessert.

El Jimador Reposado and El Jimador Blanco are now available in the export market.

EL JIMADOR

COMPANY AND HISTORY

Named after the farmers who harvest the agave plant, El Jimador is the second brand produced by the

Herradura company. Background information is therefore the same as that given above, since El Jimador is nothing more than another brand name. All of the same production methods used in making the Herradura brand products, including the use of estate-grown 100 percent blue agave, brick-oven cooking of the agave and the use of only natural yeasts in fermentation, are also used to make tequila labeled El Jimador.

Until recently, El Jimador was only sold in Mexico while Herradura was being sold in both Mexico and in the export market. However, with the increasing popularity of tequila throughout the world, the Herradura company decided to export El Jimador introducing it as a more value-oriented brand. Presently there are two El Jimador products available.

EL JIMADOR BLANCO

Type: 100 percent blue agave
Aging: **Blanco;** no aging

Unlike Herradura Silver, El Jimador Blanco is bottled without any aging. Despite their shared bottling environs, tasters preferred the Herradura Silver to the El Jimador, which was noticeably less well balanced, with a more prominent alcohol profile.

EL JIMADOR REPOSADO

Type: 100 percent blue agave
Aging: **Reposado;** aged three to four months

El Jimador's *reposado* does not really parallel any one Herradura product as it is only aged for four months. While four months is longer than the legally required minimum of two months, it is significantly less than Herradura *reposado*, which ages in wood for about a year. So El Jimador's aging is more typical of *reposados* in general. Tatsers found both of the El Jimador tequilas to be a bit harsher in the nose than their Herradura cousins and slightly less smooth. But for those who prefer Herradura's signature woody character in more modest doses, especially in light of near identical production processes, El Jimador Reposado, which is light, pleasant and by no means overpowering, is an excellent alternative, especially at its lower price point.

JOSE CUERVO

COMPANY AND HISTORY

Jose Cuervo is perhaps the best known tequila producer in the world. They are also the largest, accounting for nearly half of all of the tequila imported into the United States—more than three times that of their closest competitor, Sauza.

The company can produce documents dating back to the mid-1700s proving that it is also the oldest tequila producer still in operation, having celebrated their 200th anniversary in 1995. To hear them tell it, the early history of Jose Cuervo is in many ways synonymous with the history of tequila itself.

As the story goes, the lives of the Cuervo family and the town of Tequila began to intertwine when Don Jose Antonio de Cuervo received a land grant there from King Charles of Spain in 1758, before Mexico had become an independent republic. The land was abundant with wild blue agave, so naturally, the Cuervos began making the product then known as *vino mezcal,* which was the forerunner of what we now know as tequila.

Still, 1995 and not 1958 is considered the 200th anniversary of the company. It wasn't until 1795 that the second Jose Cuervo, Don Jose Maria Guadalupe de Cuervo, won the first exclusive official permit from the Mexican government to produce mezcal in the district of Tequila and built a distillery on the outskirts of town.

His daughter inherited the distillery and married Vincent Albino Rojas, who renamed it after himself, La Rojeña. This was not an unusual practice, and in his case it was probably deserved, as he worked so aggressively to promote the family product beyond the state of Jalisco to the far reaches of Mexico.

Though the distillery remained in the hands of Cuervo descendants, it changed names several times until the beginning of the twentieth century, when Jose Cuervo Labastida brought a modern approach to the business. It was around this time that Cuervo-produced tequila began carrying the family name.

The distillery prospered under Jose Cuervo's leadership, and the reputation of Cuervo tequila spread. After the death of Jose and his wife Ana, the business passed to

In spite of its size, many things are still done the old-fashioned way at Cuervo's La Rojeña factory in Tequila, Jalisco. Here a worker loads* piñas *by hand into a traditional* hornos *where they will be cooked. After cooking, they are unloaded onto conveyor belts, which trasnport them to the milling area.

Guillermo Freytag Shrier, and then to his son, Guillermo Freytag Gallardo, and then back to Cuervo heir, Juan Beckman Gallardo, the father of the current chairman of the company, Juan Beckman Vidal. Got that? After all of those maneuverings, the famed La Rojeña distillery, which has seen countless expansions and improvements, is still located just a few blocks from the center of the town of Tequila and now produces more than 180,000 gallons of tequila every week. Heublein, Inc., the Connecticut-based distributor, has been the exclusive importer of Jose Cuervo tequilas into the United States since 1967. And it is with its size and resources that Heublein has been able to aggressively market the Cuervo brand name and ensure a steady supply of product from Mexico to the shelves of local bars and liquor stores everywhere.

PRODUCTION AND PRODUCTS

The enormous output of Cuervo's distilleries, La Rojeña and Casa Cuervo, does not exactly allow for "handcrafting" of its tequilas. On the other hand, over 200 years of practice has contributed to what is now a mostly consistent, readily available, and reliable product line. With the company's supply of blue agave from its own as well as leased fields, Jose Cuervo produces a wide range of tequilas for every taste and budget.

JOSE CUERVO WHITE

Type: Mixed; minimum 51 percent blue agave
Aging: **Blanco;** no aging

Cuervo White is the company's standard, mixed, unaged tequila. Like Cuervo's better known Especial (described below), this tequila is made from a blend of blue agave and other sugars meeting the necessary requirement of 51 percent blue agave. Distilled to a very high alcohol content, the tequila is shipped "in bulk," mostly to the United States, and is diluted to 80 proof and finally bottled. While there are certainly many mixed Blancos available in the market, which are inferior to Cuervo White, this tequila is only recommended for serving in mixed drinks.

Jose Cuervo's current product lineup includes, from left to right, Jose Cuervo White, Cuervo Especial, Cuervo 1800, Cuervo Tradicional, Reserva Antigua 1800 Añejo, Jose Cuervo Reserva de la Familia and Jose Cuervo Mistico.

CUERVO ESPECIAL ("CUERVO GOLD")

Type: Mixed; minimum 51 percent blue agave
Aging: **Joven;** minimal or no aging

Better known as "Cuervo Gold," this ubiquitous brand retains its standing as the world's number one selling tequila. It is a *joven*, or young, mixed tequila, which according to Cuervo spends a brief period in extremely large oak tanks and is then blended with more clear tequila. Most of Cuervo Gold's color, though, like all *joven abocado* tequilas, comes from the addition of caramel coloring before bottling. Anybody who has tried tequila has probably started with a shot or two of Cuervo Gold. If, however, that has been the limit of your tequila experience, I would definitely suggest further exploration with some of the other brands described in this book for a greater appreciation of tequila's true spirit.

CUERVO 1800

Type: Mixed; minimum 51 percent blue agave
Aging: No designation given (see text)

Another very popular product for Cuervo, this one is positioned as "one step up" from Cuervo Gold. While the company characterizes 1800 as a reposado, in fact, the label bears no clear statement of an age designation.

Instead of labeling the tequila a *reposado* or an *añejo*, the description reads "a marriage of añejo and other fine tequilas." That statement implies that unaged and slightly aged tequilas have been blended together, and that not one or any other guideline for a single age designation has been met. Apparently color has also been added, as 1800 has a darker hue than many true *añejo* tequilas. Cuervo 1800, packaged in a distinctive pyramid-shaped bottle, the top of which doubles as a shot glass, is best consumed in Margaritas and other mixed drinks.

CUERVO TRADICIONAL

Type: 100 percent blue agave
Aging: **Reposado;** aged at least two months

Maybe Cuervo's least well-known and most underappreciated product, Cuervo Tradicional is a true *reposado* and is made 100 percent from the sugars of the blue agave with nothing else added. Produced in smaller quantities than the better known sister products at La Rojeña, after distillation, Tradicional rests in large oak tanks where it takes on a slight color as it mellows a bit. With its brief aging, Tradicional retains its silver hue and lively, peppery, true tequila taste. For an appreciation of pure tequila characteristics, Tradicional is the best Cuervo product for drinking straight at room temperature, chilled or on the rocks. This is the only Cuervo tequila that still proudly bears the company's original black crow (*cuervo* in Spanish) logo on the bottle neck's label.

RESERVA ANTIGUA 1800 AÑEJO

Type: 100 percent blue agave
Aging: **Añejo;** aged at least one year

In response to the growth of the premium sector of the tequila market—particularly aged products—Jose Cuervo has introduced this new product, designated as an *añejo* and as a 100 percent blue agave tequila. Utilizing the 1800 name to capitalize on its existing brand awareness, the new Reserva Antigua 1800 Añejo comes in a similar but distinct bottle, this one smaller, with a wood stopper, an agave plant etched in the glass and a gold embossed seal. As an *añejo*, Reserva Antigua 1800 shares the characteristics and nuances of taste with other *añejos*, including a greater smoothness, complexity and woody taste that come from at least one year of aging in small, charred American oak barrels.

With its 100 percent blue agave content and official *añejo* designation, this product shares little else with its predecessor, alongside which it will continue to be sold. With the increasing segmentation in the tequila market and the greater emphasis on premium products, I would not be surprised to eventually see the regular 1800 drop the Cuervo name in an effort to create separate brand identities marketed as low, middle and premium brands. Meanwhile, it is good to see the world's largest tequila company responding to the demand for better made, premium products.

JOSE CUERVO RESERVA DE LA FAMILIA

Type: 100 percent blue agave
Aging: **Añejo;** aged three years

Cuervo's most expensive product, Reserva de la Familia, was introduced in 1995 in celebration of the company's 200th anniversary. This product was created in an effort to serve the growing market for premium-aged or "ultra-aged" tequilas. With this product, Cuervo is attempting to prove that despite their size and emphasis on bulk tequila production, they can also compete with the smaller, premium-only tequila producers by making a unique product. Made from 100 percent blue agave, according to the company, Reserva de la Familia is aged for three years in new American white oak barrels. The white oak barrels have been internally charred, allowing for an exaggerated effect on the spirit from the interaction between the tequila and the wood. The result is a darker, far richer beverage with characteristics more typical of a cognac or bourbon, but with the flavor of agave present. Tasters found Reserva de la Familia to be more syrupy than some of the other ultra-aged products.

Each bottle of Reserva de la Familia is individually numbered, corked, sealed with wax, which is embossed with the Cuervo name, and then packaged in a wood box—Cuervo's version of a boutique, handmade tequila.

MISTICO

Type: Mixed—at least 51 percent blue agave
Aging: No aging

This product was really created by Cuervo for nontequila drinkers. It is a citrus-flavored tequila—basically Jose Cuervo *Blanco* with some citrus flavoring thrown in.

PATRÓN

COMPANY AND HISTORY

In contrast to tequila producing companies and tequila brands that have long histories steeped in Mexican culture, the Patrón brand's immense popularity in the premium tequila segment is something of a unique phenomenon. Patrón has managed to take the market by storm with an image based not only on quality with a fine tequila, but just as much on style, which in this case is mostly defined by its distinctive bottle.

While there is no actual distillery named Patrón, the product was until recently being made exclusively at the distillery called Tequila Siete Leguas, which is located in the town of Atotonlico el Alto in the Highlands or Los Altos region of Jalisco, east of Guadalajara. While Siete Leguas produces a very highly regarded tequila under its own brand name for the domestic market in Mexico, over half of that distillery's production was being exported as Patrón to meet the huge demand for that product in the United States.

The Patrón brand is actually owned by the St. Maarten Spirits company—which is owned by Martin Crowley and John Paul de Joria of the Paul Mitchell hair care products company. The brand's popularity would seem to be in no small part due to the fact that distribution and marketing of the product was until very recently handled by a division of the liquor giant, Seagram company.

With increasing demands on their production, in 1997 Siete Leguas was no longer able to meet the growing Patrón production demands while also producing tequila with its own name for its own market at the same time. Patrón had no choice but to find a new home. With that eventuality in mind, Seagram, which

Seagram's has just built a brand-new facility in Arandas, Jalisco, where Patrón will be manufactured.

will continue to be contractually responsible for Patrón production, planned ahead. Around the same time that I was researching this book, the Seagram company was just completeing construction of a brand-new tequila production facility located in Arandas, just up the road from where Siete Leguas is located. As no tequila has yet been produced, bottled and aged as the "new Patrón," it has obviously not yet been tasted. I would, however, venture to speculate that all of the same production methodologies that were implemented in the tequila's production previously will be utilized to ensure the same exceptional results.

According to Patrón, agave sources will remain the same. Los Altos, known for its deep red, iron-rich soils, will provide a fertile medium for all of the agave that will go into Patrón. The mineral-rich soil of Los Altos fosters the growth of larger agave plants that are intensely flavored and slightly sweeter than the plants grown elsewhere.

To date, there have been only two different tequilas from Patrón, a *blanco*, which they label silver, and an *añejo*. Both come packaged in the same handblown glass bottles with a textured surface and clear label. Once empty, the bottles have become a sought-after collectible as people have adopted them for use as water decanters or vases. In all likelihood, the unique bottle has been a significant contributing factor along with the tequila's quality for the brand's immense popularity.

The only drawback to Patrón is that without their own distillery to produce it in, there has been great inconsistency with respect to supply. Because of the lag time in growing and—where applicable—aging tequila, it will be a while before availability returns to normal and supply is able to meet demand. While an inability to meet the demand of a product could otherwise prove detrimental to that brand's image and popularity, Patrón has managed to bolster its allure. The product is not only stylishly packaged and of excellent quality, but its scarcity is almost intriguing. Once production at the new factory is in full swing, and the aging cycle catches up to demand, I am certain that the Patrón supply problems will be resolved.

PRODUCTION AND PRODUCTS

The descriptions here come with the caveat that the next generation of Patrón will be produced in a new facility making it difficult to make generalizations about production methods, style and quality. However, I would venture to guess that with the expectations that the market has, based on Patrón's reputation, the company will make every effort to ensure that the product lives up to the hype.

PATRÓN SILVER

Type: 100 percent blue agave
Aging: **Blanco;** no aging

With no aging at all, this tequila is completely clear and fairly pleasant. This product has always been, and will continue to be, made from 100 percent blue agave, the taste of which is very evident. While Patrón Silver is quite refreshing sipped straight for a true citrus fruit-agave experience, the flavors are also strong enough to hold up in a well-made Margarita or other tequila-based cocktail.

Patrón owes some of their success to the popularity of their bottles which have become collectibles. Pictured here are Patrón Silver and Patrón Añejo.

PATRÓN AÑEJO

Type: 100 percent blue agave

Aging: **Añejo;** aged at least one year

Patrón describes its *añejo* as being "a blend of three uniquely aged tequilas." Beginning with the Patrón Silver, the same tequila is then aged in small white oak barrels producing a very smooth tequila, which is probably best appreciated by sipping from a snifter.

Patrón's popularity, coupled with the raging popularity of Margaritas, has led to the popularization of that drink being made not only with silver but also with Patrón Añejo. The result is a cocktail with more body and a layered flavor profile. In general, tasters far preferred Patrón's Añejo over its silver, finding greater balance between the aroma and taste.

PATRÓN XO CAFÉ

Type: Not applicable
Aging: Not applicable

How many trends can you bundle into one product? Patrón's newest product seizes on the popularity of premium tequilas, coffee in general, and in this case specifically, coffee liqueurs. Positioned as an alternative to traditional coffee liqueurs, this product has only a 33 percent sugar content versus the higher 49 percent typical sugar content. The result is a liqueur that is dryer in flavor, lighter and less syrupy or overly sweet. XO Café is also slightly less alcoholic than tequila at 70 proof or 35 percent alcohol by volume.

Personally, I prefer to drink my coffee and tequila separately, but if I were going to drink the new Patrón XO Café it would be neat, on the rocks or in an after-dinner cocktail.

PORFIDIO

COMPANY AND HISTORY

Porfidio has attracted quite a bit of attention from the casual as well as the more informed observers of the tequila scene—despite, and maybe even due to, its rather low levels of production.

Unlike the great old tequila companies whose history is in their generations-old Mexican families, Porfidio was founded by a non-Mexican, the thirty-something Austrian entrepreneur Martin Grassl. Since Mr. Grassl founded this premium-only tequila company, Porfidio has quickly elevated itself to the top shelf of the tequila market, where it shares space with the other premium-only producers.

On its way to the top, Mr. Grassl's Porfidio brand has garnered a great deal of attention. Much of the focus is positive, but some is critical with questions about its authenticity, recognizing Porfidio more as a brand name than a tequila-producing company. On the good side and most importantly, notwithstanding its higher prices, Porfidio consistently receives high praise for the quality of its tequilas. The brand also has received a lot of positive attention as a result of its innovative packaging. Each product in the Porfidio line comes in a different and memorable bottle, ranging from one that is acid-etched blue glass to one that is handblown, beaker glass.

On the more controversial side is Porfidio's lack of a home. On the label of each bottle of Porfidio tequila appear the words, "bottled…by Destileria Porfidio." But this statement may be misleading. Destileria Porfidio is more accurately the name of a company rather than an actual place. As Porfidio tells the story, the company rents space in the factories of other tequila companies with excess capacity, takes over the production facility with its own equipment and personnel, applies its own production techniques, and produces its own unique tequilas distinctly for the Porfidio brand. All of the production has always taken place in distilleries in the Tequila area.

The mobile distillery approach must be working because despite the wonder of how quality can be maintained in such a seemingly temporary production context, the brand's many fans consistently praise the finished product for its quality. During my own travels in Jalisco, I had trouble finding a distillery where the product was actually being made for or by Porfidio.

PRODUCTION AND PRODUCTS

Since I was unable to observe Porfidio production firsthand, I have relied on the company and its representatives for an explanation of Porfidio's path from the agave fields to the bottle.

The company's philosophy begins with its stringent agave source selection with respect to proper maturity. Secondly, the use of juice only from the "first pressing" of the cooked agave is a strict requirement described by the company. Next, they utilize only natural fermentation yeasts, and as with most of the other premium tequila producers, they prefer alambic-style stills, rather than the more modern column stills. Finally, Porfidio distills to 40 percent alcohol by volume—which is its final usable

80 proof product—rather than distilling to a higher proof and than diluting the product with demineralized water, a procedure that is utilized when making lesser quality tequilas.

In a relatively short period of time, Porfidio tequilas have carved out a place alongside the other great connoisseurs' products. In fact, over half of all Porfidio is consumed in Mexico, with additional distribution not only in the United States but as far abroad as the Far East, Europe and even into the Czech Republic.

Where one would expect to typically see three different products from a single producer under the heading of 100 percent agave tequilas, Porfidio has developed a rather extensive line of fine products, each with its own defining characteristics.

PORFIDIO SILVER

Type: 100 percent blue agave
Aging: **Blanco;** no aging

Like all Porfidio tequilas, this one is made only from 100 percent agave juice. Porfidio Silver is an unaged tequila as is typical of the *blanco* category, and it is Porfidio's purest expression of the essence of the blue agave. Tasters found Porfidio Silver to be "citrusy," with one person even being reminded of "Absolut Citron."

PORFIDIO PLATA "TRIPLE DISTILLED"

Type: 100 percent blue agave
Aging: **Blanco;** no aging

The unusual use of the "plata" label alongside a separate and distinct product labeled "silver" may be puzzling, but in fact, to go along with the Porfidio Silver, this is a second unaged tequila from the same "producer." While most tequilas are distilled twice as required by government regulations, according to the company, this particular tequila goes through a third distillation. This extra distillation is intended to make the finished product smoother and even "purer" than a typical *blanco*, removing even more of the harsh "superior" alcohols. Tasters preferring the "Plata Triple Distilled" to the "Silver" commented on its smoothness. Still others preferred the Silver for its lighter, fresher and fruitier agave flavor.

The bottles used for the "Triple Distilled" may be the most beautiful of any I have seen for tequila. Crafted in Guadalajara, each is dipped in acid to achieve a frosted look. Then color is rubbed in with a cloth, starting with the deep royal blue at the base, fading to an aqua green at the top.

PORFIDIO REPOSADO

Type: 100 percent blue agave
Aging: **Reposado;** aged eight months

This new product filled the only void in the Porfidio line with a *reposado* tequila that is aged eight months in small, 100-liter, heavily toasted American oak barrels. The various barrels of tequila are blended after aging to achieve the desired balance among an authentic agave aroma and a taste with a spicy hint of wood. The amber-colored *reposado* was the most popular Porfidio tequila among tasters, who thought it very "clean" for a *reposado* with excellent clarity and flavors.

In the Porfidio tradition, the *reposado* comes in a most distinctive package. This one is round stoneware with a royal blue, porcelain finish and 18K gold lettering, with a cork to close the bottle. It is definitely the most expensive *reposado* I have seen, and presently only 2,000 cases are being produced each year.

PORFIDIO AÑEJO

Type: 100 percent blue agave
Aging: **Añejo;** aged at least two years

Another very limited production item with only 2,200 cases reaching the United States each year. Porfidio blends this *añejo* from various barrels, which are aged two and three years in previously used American oak bourbon barrels. This smooth, brandy-like tequila is very mellow, but also spicy and complex, with

Porfidio has won several awards for its unique packaging. From left to right: Porfidio Añejo Single Barrel, Porfidio Añejo, Porfidio Plata "Triple Distilled," Porfidio Silver and Porfidio Reposado.

nice flowery characteristics still evident through the wood-imparted flavors.

PORFIDIO AÑEJO "SINGLE BARREL" (CACTUS BOTTLE)

Type: 100 percent blue agave
Aging: **Añejo;** aged a minimum of one year

The "Single Barrel Añejo," known as just "Cactus," is nearly twice as expensive as the regular Porfidio Añejo, and after the introduction of Porfidio "Barrique," is now the second most-expensive tequila. The "single barrel" designation refers to the fact that rather than being blended from different barrels of aged tequilas, each bottle of "Cactus" comes out of a single selected barrel, which has been aged for somewhere between one and three years. Because of the use of tequila from a single barrel for each bottle, the color of the product may actually vary with each bottle. Only new and medium toasted, 200-liter American oak barrels are each used once in the aging process for "Cactus." The presence of new oak

Porfidio is the only company that bottles two different 100% agave tequilas. On the left is the Porfidio Plata "Triple Distilled" and on the right, the Porfidio Silver.

is noticeable, with a sweetish, vanilla-like overtone present. Tasters did not feel that the "Single Barrel" *añejo* was worth the additional cost over the regular Porfidio Añejo. With very similar aromas, the "Cactus" had a slightly sweeter, vanilla, almost cream soda-like nose with less evidence of agave flavor.

The "Cactus" bottles, which have received many awards, are actually handblown in Guadalajara. The outer bottle is made from recycled Coca-Cola® bottles, and the inside freestanding cactus "statue" is made from recycled 7-Up® bottles. Because each bottle for "Cactus" is handblown, the "fill levels" can vary with the precise dimensions of each bottle. Of course, each bottle is sure to contain the same 750 milliliters.

PORFIDIO "BARRIQUE"

Type: 100 percent blue agave
Aging: **Añejo;** aged "several" years

Not to be outdone by the competition, Porfidio has answered the call for an ultra-aged premium product with "Barrique"—which is perhaps one of the most expensive spirits in the world. Just being prepared for release at the time of this book's publication, this *añejo* is aged for what the company calls "several" years in 100-liter French Limousin oak barrels, which are called *barriques.* These are the same barrels that are used for aging

Liquid gold? This is a miniature sample of Porfidio Barrique. In addition to a fine tequila, the package is made of glass used to make laboratory beakers. A full-size bottle of this tequila is expected to retail for $500.

fine wines, such as a first-growth Bordeaux chateau. A longer aging period yields a dark, amber-like color and a powerful, cognanc-like intensity, with sweet oak flavors from the French oak.

To distinguish this product, Porfidio chose a tall, thin clear glass bottle with a small mouth and a clear, free-standing hollow cactus inside. The bottle is made from the same glass used to make laboratory beakers, making it especially strong. The gold lettering and painted-on sun add style to the beautiful package.

SAUZA

COMPANY AND HISTORY

As one of the "big two" tequila companies (Jose Cuervo being the other one), Sauza is incredibly important—not only because of its size, but also because of the importance of the Sauza family in the tequila industry past and present. The Sauza company played an essential role in establishing the roots of the tequila industry, setting the stage for, and participating in, its accelerated international growth. While Cuervo may have been the first operating tequila company, Sauza maintains the distinction of being the first to export (legally anyway) to the United States, when in 1873 three bottles of *vino mezcal* were exported from Mexico to New Mexico.

Don Cenobio Sauza fouded the company in 1873, acquiring one of Tequila's first known distilleries. This structure, located within the Hacienda de Cuisillos, was an estate owned by Mr. Pedro Sanchez de Tagle, who is widely recognized as the father of tequila. For the village of Tequila, the industry began to expand during the nineteenth century, giving rise to an elite group of entrepreneurial haciendados, among which Mr. Cenobio Sauza was the most prominent.

The distillery itself, which was then known as La Antigua Cruz ("The Old Cross") is where Don Cenobio began the production of tequila in modest amounts. In 1888, during an early symbolic effort, Don Cenobio changed the name of the factory to La Perseverancia as a reflection of the company's determination to become a force with its tequila production. Around the same time, Sauza exported its first bottle of tequila to New Mexico where the tequila won an exposition, laying the groundwork for Sauza's eventual international leadership role in the growth of tequila's popularity.

Don Cenobio is regarded as an important innovator in early tequila production, credited with introducing the impressive technical advancement of using indirect heating of the stills with steam coils rather than by direct fire. The job of continuing the growth of not only the company, but the industry as a whole, was passed on to Cenobio's son, Eladio. Born in in Tequila in 1883, Don Eladio continued the expansion of the Sauza brands throughout Mexico and internationally during a very difficult period that spanned the Mexican Revolution. Don Eladio also began to modernize the distillery by adding new equipment and machinery. Next in line was Don Francisco Javier Sauza, Don Eladio's son, who in 1931 took over the business. He continued the family's legacy by broadening distribution and working to expand Sauza's image and reputation as a producer of fine tequilas and as a leader in the industry.

With demand for tequila increasing in the 1970s, a partnership was formed between Sauza and the leading Mexican brandy producer, Pedro Domecq. This partnership led to Pedro Domecq's complete purchase of Sauza in 1988. In 1994, Pedro Domecq was acquired by Allied Lyons, and the merger resulted in the formation of the Allied Domecq company, which controls the brand today.

Sauza continues to be a major force in the tequila industry as the company continues manufacturing and distributing both bulk, mixed products, as well as 100 percent agave tequilas. While Cuervo dominates the export market, shipping approximately three times as much tequila for export as Sauza, Sauza continues to lead in domestic distribution in the Mexican market while also maintaining a high export profile. According to officials at Sauza, the company intends to continue focusing first on the domestic market—under the belief that their foundation there is most important to them.

With the international marketing power of Allied Domecq, the continuing evolution of their product line-up and the continued emphasis on quality and value, along with volume, Sauza should maintain its position as a leader in the tequila industry.

PRODUCTION AND PRODUCTS

At La Perseverancia, Sauza's factory in Tequila, the company manufactures over two and a half-million liters of tequila each year in a very high-tech, modern, pristine environment. While everything at Sauza is done on a large scale, it is also done with great care in the presence of engineers, with an emphasis on quality and consistency. The mixed tequilas and 100 percent blue agave tequilas are made in separate batches. Fermentation takes place in 75,000-liter tanks, and then the *mosto* is pumped to the distillation area.

One of the highlights of a visit to La Perseverancia is the mural in the entry courtyard, which has been reproduced in so many places and has become emblematic of tequila and its history. Painted in 1969 by the Jaliscan muralist Gabriel Flores, the painting has three distinct parts. The first scene is a portrayal of the myth of tequi-

Sauza's La Perseverancia was the only factory I visited where the agave is shredded before it is cooked. A front-end loader drops the agave into a hopper before they are transported by conveyor belt to the desgarradora.

la's origin. The second scene is an illustration of tequila's rudimentary production process. The third and largest scene depicts the joy caused by the excessive consumption of tequila. At the center of the mural is a rooster standing on top of an agave plant, symbolizing the nobleness and courage of the Sauza brand. (see reproduction on page 11.)

The first step in the production process utilized at Sauza—which is different from any that I saw in the many other distilleries I visited—is that rather than milling the agave after cooking, the raw agave is instead loaded onto a conveyor belt and fed into a "tearing machine" called a *desgarradora.* There the agave is shredded before it is placed in autoclaves where it is cooked for twelve to fourteen hours. After cooking, the cooked fibers are milled again, and the extracted juices are combined with juices that were extracted during cooking. Formulation and fermentation follow before distillation, aging and eventual bottling.

SAUZA SILVER

Type: Mixed; at least 51 percent blue agave
Aging: **Blanco;** no aging

This is Sauza's workhorse tequila. It is actually the largest-selling tequila in Mexico, where Sauza is still the foremost name in tequila. Made from 51 percent blue agave, after other sugars are added during fermentation, Sauza silver is distilled to 55 percent alcohol by volume and then sent to diluting tanks where demineralized water is added to bring it down to the required 40 percent alcohol before bottling. In the United States, this blended tequila would typically be found in "wells" of discriminating restauarants for use in their mixed drinks.

SAUZA EXTRA

Type: Mixed–at least 51 percent blue agave
Aging: **Joven;** no aging

"Extra" is Sauza's "Gold" tequila and its answer to Cuervo's version—which is the number one selling tequila outside of Mexico. Sauza Extra is made in essentially the same way as Sauza Silver, with the addition of coloring and flavoring to give the tequila its golden hue as well as a smoother taste. Like Sauza Silver, Sauza Extra is shipped in bulk to various locations in Mexico and the United States where it is bottled. Silver and Gold account for approximately 700,000 cases of production each year.

HORNITOS

Type: 100 percent blue agave
Aging: **Reposado;** aged at least three months

For some time, Hornitos has been Sauza's most heavily exported 100 percent blue agave product. While this tequila has long been very popular in Mexico, drinkers

Sauza Hornitos rests in large wooden tanks for approximately three months before bottling.

At Quinta Sauza, the house bartender demonstrates the proper way to make a "Fresca" using Sauza Hornitos.

in the United States are taking an increased liking to this tequila as they gain an appreciation for the *reposado* category of tequilas in general. After distillation, Hornitos is aged in large, 40,000-liter wooden tanks. While aging in wood, Hornitos softens a bit and naturally takes on a slight yellow color, but does not adapt what is known as a "wooded personality."

With its 100 percent agave content, tasters favored Hornitos' "excellent agave character," while not being as "harsh" as most silver tequilas as well as many *reposados.* While Hornitos, with its versatility, can be enjoyed straight or in mixed drinks, I particularly enjoyed it at Sauza mixed with limon (a Mexican verson of lime, but smaller) and Squirt—which is the locally made grapefruit soda in Tequila. Many tasters noted how well they thought Hornitos would stand up to citrus mixes or fruit juices in general. Hornitos is one of my favorite tequilas for making Margaritas. In fact, while sampling several tequilas straight, one taster remarked that "[Hornitos] makes me really want a Margarita." That comment is a testament to Hornitos' crisp, fruity, light character. Fortunately, because of Sauza's brand presence, Hornitos is the most widely available of the premium *reposados*. Considering quality alone, but particularly with respect to price and value, Hornitos continues to stand out.

SAUZA CONMEMORATIVO

Type: Mixed; at least 51 percent agave
Aging: **Añejo;** aged at least one year

Sauza's very popular Conmemorativo is one of the few mixed tequilas on the market that is aged long enough to be designated as an *añejo*. The result is a product that has a lot of character and offers a lot of quality for the price—roughly one-third less than the 100 percent *añejo*. bottlings. In fact, Conmemorativo is a testament to the fact that a quality product can in fact be "blended." At La Perseverancia, the Sauza factory in Tequila, Conmemorativo is aged for between a year and a year and a half in small white oak barrels.

Packaged in a distinctive brown glass, high-shouldered bottle, Conmemorativo has become notorious for its use in quality Margaritas. When adding flavors to a Margarita, the body and smoothness of the resulting drink's taste—together with the agave flavor—is very important. At Mesa Grill, our signature "Cactus Pear Margarita" has long been made using Conmemorativo for just that reason.

GALARDON

Type: 100 percent blue agave
Aging: **Reposado;** aged eleven months

While techically this tequila is categorized as a *reposado*, for the brand-new "Galardon," Sauza has created the unofficial classification *Grand Reposado*. Because most *reposados* are typically aged for between three and nine months, Sauza added the "*Grand*" to indicate a tequila that has been aged for eleven months. Literally translated, *galardon* means "the highest prize."

With the increasing popularity of premium tequilas, to make Galardon, Sauza is using only juices fermented

from 100 percent blue agave to make this special tequila in very, very small quantities. The extended aging time, which takes place in small white oak barrels rather than the larger tanks in which Hornitos is aged—gives Galardon a more amber-like color and a smoother texture than Hornitos. Sauza has also created a brand-new package with a metal label meant to represent the craftsmanship of the traditional Mexican artisan. Additionally, each bottle is signed and numbered. If you are able to find a bottle of Galardon, consider yourself lucky as only 2,000 cases of this product are expected to be exported each year.

Tasters were especially impressed with Galardon. To quote two tasters: "I love this. I could drink it all night long." And, "Would be great to drink after dinner...a smooth finish." Apparently, Sauza succeeded in its goal of creating a tequila which combines the freshness and clarity of a *reposado* with the texture of an *añejo.*

SAUZA TRES GENERACIONES

Type: 100 percent blue agave
Aging: **Añejo;** aged at least two years

Sauza's "top of the line" tequila is nicknamed "Three G's." Tres Generaciones comes in a black-etched bottle embossed with portraits of the three generations of the Sauza family who founded and built the brand up to its present status as a giant in the tequila world. In celebration of the Sauza family, Tres Generaciones is Sauza's most expensive product, made only from 100 percent blue agave and, according to the company, aged for at least two years in small white oak barrels.

Tasters liked Tres Generaciones' nice aroma and hon-eyed-like flavor reminiscent of "crème brûleé." They also commented on the tequila's "good, smooth, nice mouth feel," yet still found the tequila to be "soft and light"

Sauza products at top left to right—Sauza Conmemorativo, Sauza Hornitos, Sauza Extra, Sauza Silver. At bottom—Tres Generaciones and the limited Galardon.

enough to be eminently enjoyable. Three G's is definitely Sauza's smoothest tequila and the one best suited for sipping and savoring.

TRES MUJERES

COMPANY AND HISTORY

Of all of the tequila brands that I chose to write about for this book, Tres Mujeres is probably the least well-known. In fact, with no prior knowledge of its existence, I stumbled on this company's *fabrica* only to find that distribution in the United States was imminent. For certain, Tres Mujeres' notoriety bears no relationship to the quality of the tequila being produced. With the tequila just beginning to appear in the United States markets now, I expect that its quality will soon make it popular among informed tequila drinkers.

The Tres Mujeres brand is actually the product of a company named after its owner, J. Jesus Partida Melendrez. While the Tres Mujeres brand is only two years old, its producer is no stranger to the world of blue agave. The Melendrez family has been growing millions of their own blue agave plants for other more well-known tequila producers for over sixty-five years.

As the better quality tequilas became more popular, Jesus Partida decided to enter the fray and challenged himself to make the best 100 percent blue agave, 100 percent naturally produced tequila that would compete with the other premium producers, but at a slightly lower price point. Thus was born Tequila Tres Mujeres, which translated literally is "Three Ladies"—in honor of Jesus' mother and two sisters.

PRODUCTION AND PRODUCTS

Driving down a highway in the town of Arenal, which is west of Guadalajara and just near Amatitán in the valley, I passed a roadside stand where Tequila Tres Mujeres was being sold. Behind the stand and down a bumpy dirt road, past the Partida agave fields, is a rather small building which is where Tres Mujeres is manufactured using a traditional approach.

Tres Mujeres Blanco and Tres Mujeres Reposado are packaged in identical etched bottles.

All of the blue agave is handcut and baked in one small oven for twenty hours and then left to rest for another twenty-four hours before the oven is opened. The juices are extracted from the cooked agave as they are passed through a shredding machine while water is added. Natural fermentation occurs in three modestly sized, stainless steel fermentation tanks over an eight- to ten-day period with no artificial or chemical yeasts added. The first and second distillations are then done in two pot stills. Only 1,000 liters of Tres Mujeres tequila are produced every day by a total of less than twenty people working at the factory.

TEQUILA TRES MUJERES BLANCO

Type: 100 percent blue agave
Aging: **Blanco;** no aging

As described above, this tequila is "all natural," and it comes in a very attractive, etched glass bottle. While my visit to the small but impressive Partida facility made me want to write about Tres Mujeres, it was during a comparative tequila tasting back home in New York at Mesa Grill

The Tres Mujeres Reposado Anfora bottling contains the same tequila as the regular reposado but comes in a special, leather-trimmed, canteen-style package.

that I became more excited about the brand. All of the tasters were impressed by this tequila which none of them had ever heard of before. The *blanco* was incredibly vibrant and flavorful yet smooth. "Spicy, chile pepper" and "nutty" were descriptions attributed to the *blanco*, which provides the drinker with a very pure, genuine tequila experience.

TEQUILA TRES MUJERES REPOSADO AND "REPOSADO ANFORA"

Type: 100 percent blue agave
Aging: **Reposado;** aged three months

Every bit as impressive as Tres Mujeres' *blanco* is their *reposado,* which after distillation is aged for three months in new oak barrels in a *bodega,* which is located within the facility I visited. The *reposado* is available in two different packages. The first is in the same etched glass bottle as the *blanco*, and the second is in an eye-catching printed faux leather *anfora* (canteen) bottle with a strap and leather trim.

While the leather casing and the leopard print on the *reposado* gift packaging may seem a bit showy for such an otherwise understated brand's image, it should be recognized as a sign of the times. With so many tequilas competing for consumers' attention, in an effort to "stand out," the package has become an important consideration for all tequila producing companies, importers and their distributors. In fact, while Tres Mujeres does not have an *añejo* available in the United States yet, it is sure to be noticed when it does arrive. During my visit at the distillery, Jesus' son, Sergio, who carries the title of "Coordinator of Special Events," showed me some of the bottles that they were experimenting with for their premium product to eventually be packaged in.

Tasters found the Tres Mujeres *reposado* to possess even more character than the *blanco*, though similar in style and with the same emphasis on agave flavor. "Peppery, lots of flavor and aggressive taste" was a description used for this honey-colored, pleasantly surprising tequila.

MEZCAL

Tequila's surging popularity, its increased distribution and the growing awareness of its premium brands have all spurred a renewed interest in mezcal. Similarly, there are signs of growing appreciation of premium mezcals as a category. With an estimated 100,000 cases of mezcal being sold in the United States annually, compared with more than 5 million cases of tequila, mezcal may seem like just a blip on the screen, but it is a blip that I expect will continue to get brighter.

Like tequila, mezcal is distilled from the agave plant family, but from different species of agave. The mezcal agave varieties are more typical of Oaxaca, the southern Mexican state where the best mezcals arc produced. The most common species of agave used to make mezcal is the *espadin.* Aside from being made from a distinct plant variety, mezcal differs from tequila in the method used in cooking the agave. Rather than baking the agave in ovens, in mezcal production the agave is roasted over charcoal, giving it distinctly smokier flavors.

At the same time that tequila has become more and more popular outside of Mexico, awareness about mezcal has increased as well. Just as in the case of tequila, there is a range of mezcals on the market from artisanal, handmade, interesting brands, to more mass-produced and less enjoyable products.

Until the late 1800s when tequila was recognized as separate and distinct from other agave-derived drinks, mezcal, or *vino mezcal* as it was called, and *pulque* were the predominant beverages of choice. Before the designation of tequila's existence as a product of place, only aristocrats were permitted to drink the forerunner, distilled mezcal, while commoners consumed *pulque* the undistilled liquor made from fermented agave pulp.

By definition, technically all tequilas are in fact mezcals but not all mezcals are tequilas. In the same way that tequila has been strictly defined by the Mexican government with respect to where and how it is made, today mezcal must meet specifically defined regulations. The relationship betwen tequila and mezcal resembles that of cognac and Armagnac. Both are pairs of similar but significantly different products, made in different areas, neither one of which is necessarily superior to the other.

REGULATING MEZCAL

While historically mezcal could be made anywhere in Mexico, recently the government has become more involved in codifying the industry. The phrase "bottled in *origin*" appears on mezcal bottles limited to the areas where they are in fact produced.

The first new regulation of mezcal production was in 1995 with the creation of a special denomination of *origin* for Oaxacan mezcal. Together, the Mexican government and the mezcal producers declared an approved geographic region for designated mezcal production—predominately in the state of Oaxaca. The regulation also permitted the production of mezcal in the states of Guerrero, Durango, San Luis Potosí and Zacatecas, which were approved as "designated origins." Therefore, mezcals produced and bottled in these designated areas can be labeled "bottled in origin," whatever that origin might be. The government's regulatory aim was to further define mezcal as a premium spirit, in turn giving the local economies in general, and the mezcal industry in particular, a boost. Mezcals that are not of an approved designated origin may actually be utilizing a product wholly or partly made in Oaxaca but bottled elsewhere. In that case, the bottle may state something like *Regional de Oaxaca* rather than stating that they are bottled in a particular origin. In these cases, the label will only broadly state that it was bottled in Mexico usually with the phrase *Hecho en Mexico.*

In 1997 the government developed a new *Norma* for mezcal establishing new requirements pertaining to mezcal production, including the creation of two separate classifications of mezcal much like the different tequila categories. This law was intended to regulate the "pureness" of mezcal. The first classification, "Type 1" refers to mezcals that are made from 100 percent agave, or *maguey* as it is more typically known in the mezcal world. The second classification, "Type 2," refers to mezcals that must be made of at least 80 percent *maguey.* As in tequila production, the balance of the content may be of various other non-agave fermentable sugars. Typically, there is a correlation between *maguey* content and the production methods used to make the mezcals in each category. Type 1 mezcals are typically more artisanally produced, manufactured in much smaller quantities and are more expensive than Type 2 mezcals. The latter are more likely mass produced using more modern, industrially advanced methods and are less expensive.

The new *Norma* also established the aging categories

for mezcal—similar to the tequila categories for *joven, reposado* and *añejo.* Like tequila, mezcal can be colored with various artificial colorings. Aging of mezcals is not nearly as common as it is with tequilas. After sampling a few *reposado* and *añejo* mezcals, I have found for my palate that the effect of aging actually diminishes, or at least disguises, the true mezcal characteristics. Aging mezcal masks its natural flavors causing them to taste more like aged tequilas. The colorants used on the more mass-produced mezcals additionally have the same obscuring or softening effect as they do on gold tequilas.

MEZCAL PRODUCTION

To illustrate the basis for mezcal's flavor profile, I will describe the traditional process of mezcal production so that its unique taste can be understood. As I have mentioned previously, there are several differences between tequila and mezcal production methods. The actual plant variety used is different from tequila, as is the cooking process of the agave after the harvest. Various species of the maguey plant are used to make mezcal including the giant *pulque maguey, maguey silvestre* (wild), *maguey tobala* (a rare variety from the mountains), *maguey espadín* (sword, the most commonly used), *maguey tepestate* (horizontal), and *maguey larga* (long), which is a larger variety of the *agave azul,* or blue agave.

Mezcal is made from *piñas* weighing between 60 and

In Oaxaca, harvested agaves are taken by burros from the fields to the palanque where mezcal is made.

In mezcal production the agave is roasted in an underground, stone-lined pit.

120 pounds, harvested from seven- to ten-year-old maguey plants. After the agave plants are harvested for mezcal production in very much the same was as they are for tequila, the *piñas* are brought to the *palenques* or factories or distilleries where they are processed. They are first unloaded and then cut up into smaller sections. Then, they are placed in a rock-lined conical pit, which is about twelve feet in diameter and about eight feet deep, capable of holding more than three tons of *piñas.* The pit is preheated with wood and covered with a bed of river stones, which prevents direct contact of the *piñas* with the charcoal. The *piñas* are then covered with hot rocks that have been heated in a wood fire. They are then covered with a layer of the leaves or fiber from the plant, followed by woven palm-fiber mats and finally a layer of earth. A *mezcalero* or *practico* oversees the roasting for two or three days, as the agave continues to absorb flavors from the earth and wood smoke. As a result of this process, the flavor characteristics of tequila and mezcal greatly diverge. Instead of following tequila's process of steaming agave in stone ovens or stainless steel autoclaves, mezcal's agave is in effect roasted, imparting its distinctly smoky flavor. During the baking process carbohydrates or starches contained in the *piñas* are converted into fermentable sugars.

After cooking, with its now caramel-like sweetness and

smoky quality, the agave is ready to be crushed or milled for the extraction of its juices. There is a range of methods used in this step that vary according to the size and style of a particular mezcal production facility. In traditional mezcal production, a *tahona*-like stone wheel is used to crush the cooked agave. In more modern mezcal-producing operations, "tearers" and other mechanical milling equipment are used to more quickly handle the same task.

The crushed agave is then placed in either wooden vats or stainless steel tanks, and water is added. The mash, which is called *tepache*, is left to ferment for anywhere between four and thirty days depending on the season. Warmer weather speeds up the fermentation. Together, the mezcal solids and liquids are then transferred to steel, copper or, in the most traditional applications, ceramic pot stills where distillation takes place. Whereas tequila production has always required two distillations to achieve the purest products, traditionally mezcal was distilled only once. More recently, however, mezcal has undergone double distillation as well. The more carefully made mezcals are distilled precisely to the alcohol content at which they are then bottled. The mass-produced mezcals are distilled to a higher proof and then are later diluted with purified water before bottling.

The differences between traditionally produced or "handmade" mezcals and the more modern versions is just as striking as the differences between the traditional and more modern tequilas. In any case, but especially in the small villages where limited-production mezcals are made, there is great reverence for mezcal and its associated traditions. It is valued for its traditional ceremonial and medicinal uses among villagers as well as for its social use.

WHO GETS THE WORM?

There are two different types of small worms that live inside the agave. Red worms live in the roots, and white worms live in the leaves. Legend has it that the worms, while living within the agave, inherit the plant's magical spirit and carry the spirit of the revered plant with them to the mezcal and, more importantly, eventually to the person who drinks it. In reality, more than anything else, the worm is a marketing tool, encouraging conversation around the ceremony of finishing each successive bottle of mezcal to see "who gets to drink the worm." Let it be known that besides adding flavor to the mezcal, the worm is harmless and is actually said to be a source of protein.

The following story, told to me by Ron Cooper of Del Maguey, Ltd. Mezcal Company, explains the origin of the worm in the bottle. In 1940 Jacobo Lozano Páez moved to Mexico City from Parras de la Fuente, Coahuila, to study art. He got a job at "La Minita" affiliated with "La Economica" and this experience changed his artistic aspirations to those of a successful bottler and trader of mezcal, an activity initiated in the same liquor store. Jacobo met his future wife working there. He started a small bottling facility in 1942 and entrusted it into his wife's hands.

The couple bought mezcal from the Méndez family in Matatlan, Oaxaca. They collected and cleaned bottles for their operation. In 1950 the inexperienced entrepreneur, now owner of Atlántida, S.A.—a small bottling company located downtown—and a connoisseur of the mezcal's production process, discovered that the maguey worms gave the mezcal a special flavor, since when the plant was cut for cooking many of these creatures remained in the heart. This is how he got the idea to give his product a distinctive touch, adding a worm to the beverage and including with the bottle a small sack with salt seasoned with the same larva, dehydrated and ground. These ingredients determined the identification of the mezcals "Gusano de Oro" and "Gusano Rojo" mezcals to the United States, and led to the inclusion of the worm in many other mass-produced and mass-marketed mezcals.

MEZCAL BRANDS

Tequila brands, however, still outnumber mezcals in the United States while mezcal is still far more accepted in Mexico than abroad. Basically the mezcal market in the United States can be divided into those products available from the large commercial producers and those from the small artisan producers.

THE BIG PRODUCERS

The more widely known commercial producers include Monte Albán and Dos Gusanos from the company Mezcal Monte Albán S.A., and Gusano Rojo from the company Nacional Vinícola, S.A. These are the mezcals

Gusano Rojo and Monte Albán are two of the mass-produced mezcals from large companies.

that are likely to come with the infamous worm in the bottle. While all of these now commercially produced mezcals were born of small family producers in the early 1900s, they all have evolved into large, modernized production operations similar to those employed by the large tequila producers. All have incorporated autoclaves for the cooking of agave hearts, the use of a mechanical shredder for juice extraction process, the use of stainless steel containers for fermentation and continuous distillation stills. These companies also similarly have alliances with the large international liquor conglomerates with wide distribution and promotional networks.

The mezcal from the large producers can be equated to the bulk products from the larger tequila producers. Typically, other non-agave sugars are added before fermentation, and coloring is added after distillation. The resulting product lacks the genuine, pure-taste character of the better tequilas, and especially that of the better mezcals. Tasters found them to possess not only the familiar smoky mezcal characteristics, but also a more alcoholic, sometimes described as a slightly "chemical" character, which may also be evident in mass-produced tequilas reserved for use in mixed drinks.

THE BOUTIQUE PRODUCERS

Encantado and Del Maguey Single Village are examples of two smaller mezcal brands from small producers that have begun importing their premium mezcals to the United States. These mezcals are beginning to appear in the best liquor stores and on the bar shelves of fine restaurants. Their appealing products and fascinating histories provide colorful examples of the growing premium sector of the mezcal industry.

ENCANTADO MEZCAL

Encantado, which translated literally means enchanted, was the first superpremium mezcal to be made available commercially in the United States when it was introduced in 1995. The brand is the creation of the team of Carl Doumani, the well-known vintner of the Stags Leap Winery in the Napa Valley, and Pam Hunter, a Napa Valley-based publicist who also happens to be Doumani's frequent traveling partner in Mexico. The goal with Encantado was to create a true artisanal product to parallel the finest handcrafted tequilas.

In their quest for not only excellence, but also authenticity, Doumani and Hunter sought out the assistance of several of Mexico's most esteemed experts in maguey cultivation and mezcal production. Together they identified what they believed to be Oaxaca's finest maguey plantations. These areas cultivated four varieties of maguey—the *Espaidín*, which is a blue variety, as well as two cultivated types, *Karwinski* and *Americana*, and one wild agave called *Tobala*. All four agave varieties are cultivated to produce Encantado's complex yet soft, smoky flavor. Encantado utilizes the most primitive and least interventionist methods in the making of its 100 percent agave mezcal, similar to the more traditional tequila producers.

The agave used by Encantado is farmed on various

Crafted the traditional way, Encantado was the first handmade mezcal to be exported.

very small, family-owned ranchos. After harvesting the *piñas*, they are actually brought by burro to village distilleries where they are roasted with wood in large, outdoor, stone-lined pits, as described above. After four days of roasting, the *piñas* are transported in small wheelbarrows to a nearby stone *molino* where they are ground before fermentation. Fermentation then takes place for about two weeks in wooden vats. The fermented juice is then transferred to stills made of clay called *ollas* for their first distillation. The second, and final distillation, then occurs at one central facility after a very careful blending of the components taken from each of Encantado's twenty-nine *palenques,* or small distilleries.

Tasters are impressed by Encantado's mezcal with its peppery and slightly spicy qualities. "More fiery" and "rustic" are typical comparisons to tequila, which tends to be much smoother in character. "Herbaceous, flowery nose" with flavors described as "herbal and earthy" define this most complex but clean beverage with an unusually long smoky finish that is not easily forgotten.

DEL MAGUEY SINGLE VILLAGE MEZCALS

Newer on the market is a collection of four distinct "single village" mezcals from small producers in Oaxaca, which are some of the most unique products I came across of any type, while researching this book. Each is a pure, organic and unblended spirit with exceptionally vivid flavors. Del Maguey was founded in 1995 by Ron Cooper, an artist from Ranchos De Taos, New Mexico, who spent considerable time working in Oaxaca when he fell in love with these unique mezcals.

The single village mezcals from this company are truly handmade and offer mezcal's answer to the literally hand-casted tequilas from El Tesoro. In the same vein as a "single vineyard wine," each of the four mezcals from Del Maguey is unblended and from a single, tiny remote village in various areas of the state of Oaxaca. Here, only natural processes are utilized as the only ingredients used by each village *palenquero* are the agave plant and water in a long, drawn-out process that captures mezcal's true spirit. In an effort to preserve the exceptional quality, production is limited to 3,200 bottles per year for each different village's mezcal.

Del Maguey uses horse-pulled stone grinders to ground the maguey into a mash, which is then put into a large oak vat for fermentation. During fermentation, only natural yeasts are used, with no addition of chemical catalyzers to speed up the process.

The resulting products are truly unique. Each has a distinct, smoky flavor, which is very potent. Unlike the mass-produced mezcals, which are distilled to a high alcohol content and then diluted with water to reach 80 proof, each of the four Del Maguey mezcals is distilled to its final alcohol content depending on its character.

Mezcal Santo Domingo Albarradas (98.4 proof), Mezcal Chichicapa (95.5 proof), Mezcal San Luis Del Rio (96.6 proof), and Mezcal Minero (98.2 proof) are all from distinct villages with their own specific environments that provide them with their own distinctive qualities.

Tasting these mezcals is an intense experience, and one must be prepared to be filled with their warmth. Each has a deep, complex, smoky flavor that the uninitiated drinker may interpret as "burnt," but which is indicative

Del Maguey Ltd. produces four distinctive mezcal, each from a different, remote village in Oaxaca.

of the cooking technique used. Beneath that upfront impression they are all spicy, sweet and possess hints of roasted tropical fruits with some citrus overtones.

My favorite Del Maguey mezcal is the Minero, with its notably floral nose with vanilla overtones leading up to a burnt-honey flavor with a bit of lemon. It is deep, warm and sweet all the way to the finish with a unique fruity and smooth taste. At the *palenque* where Minero is made, the still is made of clay with bamboo tubing rather than the usual copper stills and tubing.

The artisanal flavor of Del Maguey is further enhanced by its packaging. Each bottle comes in its own unique, traditional, handwoven, palm-fiber basket. The women of a particular area in Oaxaca have been weaving palm-fiber baskets for thousands of years, and all their designs are of ancient Zapotec or Mixtec origin. The designs are abstractions of flowers, ceramics or architectural elements, and each basket takes one woman one day to weave. The labels are also beautiful, each a drawing by artist Ken Price.

THE ULTIMATE MARGARITA

Every Margarita recipe is really a variation on the same theme, which is defined by the inclusion of tequila, orange liqueur, lime juice, ice and salt.

There are probably as many versions of the "ultimate Margarita" recipe as there are tales purporting to recount the true story of the Margarita's invention. Since this book is meant to help you become a better-informed drinker, rather than an impressive storyteller, we'll stick to an exploration of Margarita-making ingredients and techniques.

TEQUILA

Personal preferences about which brand and what type of tequila to drink straight should carry over to tequila preferences for use in a Margarita. If you want the fresh, true agave bite that you prefer when drinking a *blanco*, then use this tequila in your Margarita. It will supply real punch to your cocktail. Similarly, the added smoothness and complexity of a *reposado* or, to an even greater extent, an *añejo*, will make a Margarita darker in color but fuller-bodied, with more complex flavors.

My preference is to use a *blanco* tequila as it complements the freshness of the lime juice and orange liqueur. A *reposado*, preferably one of the lesser-aged products, also may be used for a softer or mellower Margarita. While some people do enjoy *añejo* Margaritas that contain hints of oak and vanilla, I find those to be such a departure from the bright Margarita spirit that I suggest sticking to the *añejos* for sipping straight.

The issue of whether or not to use 100 percent agave tequila is related to the purity of the Margarita. Is it worth the extra money for a "premium" Margarita if it is going to be mixed with the other ingredients? While I am an advocate of 100 percent agave tequilas, I also am realistic about their relative virtues in the recipe at hand. When you begin your education in tequilas, a good quality non-100 percent agave tequila will do fine in the cocktail. As you become more attuned to tequila's nuances, you will probably begin experimenting with different grades of tequila.

ORANGE LIQUEUR

Orange liqueur is the second most prominent ingredient in a Margarita. The choices are far fewer than tequila choices, but no less important in their impact. Triple sec is the most popular; it is basically a liqueur made from orange skins that have been fermented and then distilled into an alcoholic product. There are many different brands of triple sec, the only real differences being the alcohol levels to which they are distilled, and whether or not any additional sweeteners were added. I suggest using a triple sec that is around 30 proof, because the tequila should be contributing most of the alcohol and flavor to the Margarita, not the orange liqueur.

There are a number of premium orange liqueurs available. At Mesa Grill we offer Cointreau and Grand Marnier, the two most popular orange liqueurs. Our "Especial Margarita" is made with Gran Torres. Other products less widely available are Citronge and Controy.

Cointreau is my favorite orange liqueur, and in my opinion makes the most flavor-balanced Margarita. It is a triple-distilled, 80-proof spirit made from orange peels that have been dried out. The oil from the peels is then blended with grain neutral spirits and pure sugar cane before being distilled three times. Margaritas made with Cointreau in place of triple sec have a slightly richer orange flavor. But it is important to use Cointreau, or any liqueur, in the proper proportions so the orange flavor does not overwhelm the tequila.

Grand Marnier is a premium orange liqueur that is blended with Cognac and then aged for one or two years. Margaritas made with Grand Marnier are darker in color and take on a very distinct flavor and character. The presence of Cognac in the Grand Marnier introduces a new flavor into the Margarita, and while many people like those cocktails, I prefer the purer flavor of tequila, lime and the hint of orange which triple sec or Cointreau imparts.

LIME JUICE

The last essential Margarita ingredient—and one that prompts the least discussion—is the citrus juice. Some people prefer to squeeze lemons rather than limes because of the seasonal inconsistency of the limes' tartness. My suggestion is to buy and squeeze your own limes. If the lime juice is too tart you may add some sugar or a commercially produced sour mix. Depending on how many Margaritas you are making, you may use store-bought, freshly squeezed lime juice. Be careful, however, that you are not buying one of the commercially pre-sweetened products. Fresh is always the best.

SALT

Purely a matter of personal taste. Salt should never be mixed into a Margarita, but some people like to have the rim of the glass salted. Always use kosher or sea salt. There is nothing particularly authentic about this ingredient; rather, it is probably a result of the lime-and-salt convention used when "shooting" tequila (see p. 165). Personally, I think salt just gets in the way of the taste of the other ingredients.

NOTE: The ingredient "white" tequila, in the following drink recipes, may refer to *blanco, plata* or silver tequila, depending upon the manufacturer.

MESA GRILL MARGARITA

MESA GRILL MARGARITA

Below are the basic ingredients for making any of the Margarita variations that follow this recipe.

- 4 parts white tequila (2 oz.)
- 2 parts triple sec or other orange liqueur (1 oz.)
- 2 parts lime juice (1 oz.)
- Coarse salt (optional)
- Lime wedge

VARIATIONS:

ROCKS MARGARITA

Combine tequila, triple sec and lime juice in a cocktail shaker with ice cubes and shake well. Pour into a chilled Collins glass and garnish with the lime wedge.

Optional: Rub a lime wedge around the rim of the glass and dip the glass rim into a saucer of coarse salt.

UP MARGARITA

Combine the tequila, triple sec and lime juice in a cocktail shaker with ice and shake well. Strain into a chilled cocktail glass and garnish with the lime wedge.

Optional: Rub a lime wedge around the rim of the glass and dip the glass into a saucer of coarse salt.

FROZEN MARGARITA

Combine tequila, triple sec, lime juice and ice in a blender and blend until smooth. Pour mixture into a goblet and garnish with a lime wedge.

Optional: Rub a lime wedge around the rim of the glass and dip the glass rim into a saucer of coarse salt.

ABSOLUT TEQUILA

- 1 part white tequila ($^1/_2$ oz.)
- 1 part Absolut® vodka ($^1/_2$ oz)
- 4 parts orange juice (2 oz.)
- Lemon slice

Combine liquid ingredients in a cocktail shaker with cracked ice and shake well. Pour into a highball glass and garnish with lemon slice.

ACAPULCO

- 3 parts white tequila (1 $^1/_2$ oz.)
- 4 parts pineapple juice (2 oz.)
- Sprite® or 7-Up®

Combine tequila and pineapple juice in an Old-Fashioned glass over cracked ice. Fill to the top with Sprite® and stir.

ACAPULCO CLAM DIGGER

- 3 parts white tequila (1 $^1/_2$ oz.)
- 6 parts tomato juice (3 oz.)
- 6 parts clam juice (3 oz.)
- $^3/_4$ tablespoon horseradish
- Tabasco sauce to taste
- Worcestershire sauce to taste
- Splash of lemon juice
- Lemon or lime slice

Mix all the ingredients in a glass with cracked ice. Garnish with lemon or lime slice.

Note: Clamato juice (6 oz.) may be substituted for the tomato and clam juices.

AMORE

AMORE

- 2 parts gold tequila (1 oz.)
- 1 part orange curaçao ($^1/_2$ oz.)

Combine the tequila and curaçao over cracked ice in an Old-Fashioned glass and stir.

ARRIBA!

- 3 parts white tequila ($1^1/_2$ oz.)
- 6 parts grapefruit juice (3 oz.)
- Club soda

Combine all ingredients except soda in a cocktail shaker with cracked ice. Pour into a highball glass and fill with soda.

BERTA'S SPECIAL

- 4 parts gold tequila (2 oz.)
- 1 teaspoon honey
- 1 egg white
- 5–7 dashes orange bitters
- Juice of one lime
- Sparkling water
- Lime slice

Combine all ingredients except sparkling water and lime slice in a cocktail shaker. Shake vigorously and pour into a chilled Collins glass filled with ice cubes. Top off with sparkling water and garnish with lime slice.

BLACK DOG

- 3 parts gold tequila (1 1/2 oz.)
- Coca-Cola®

Pour tequila into an Old-Fashioned glass over ice cubes and fill with Coca-Cola®.

BLACK DOG

BLOODY TEQUILA MARIA

- 2 parts white tequila (1 oz.)
- 4 parts tomato juice (2 oz.)
- Dash lemon juice
- Dash tabasco sauce
- Dash celery salt
- Lime or celery stick

Combine liquid ingredients in a cocktail shaker with cracked ice and shake well. Strain into a chilled Collins glass over ice cubes. Garnish with lime or celery stick.

BLUE SHARK

- 1 part white tequila (1/2 oz.)
- 1 part vodka (1/2 oz.)
- 1 part blue curaçao (1/2 oz.)

Combine all ingredients in a cocktail shaker with cracked ice and shake well. Strain into a chilled shot glass.

BLUE SHARK

BLUE MOON

- 3 parts gold tequila (1 $^1/_2$ oz.)
- 6 parts orange juice (3 oz.)
- 1 part blue curaçao ($^1/_2$ oz.)

Combine tequila and orange juice in a chilled goblet over crushed ice and stir. Float curaçao on top.

BOMBA CHARGER

- 3 parts white tequila (1 $^1/_2$ oz.)
- 5 parts pineapple juice (2 $^1/_2$ oz.)
- 1 part lemonade ($^1/_2$ oz.)
- 1 part grenadine ($^1/_2$ oz.)

Combine all ingredients in a cocktail shaker with cracked ice and shake well. Strain into a chilled cocktail glass.

BOMBA CHARGER

BUNNY BONANZA

- 4 parts gold tequila (2 oz.)
- 2 parts apple brandy (1 oz.)
- 1 part lemon juice ($^1/_2$ oz.)
- $^3/_4$ teaspoon maple syrup
- 3 dashes triple sec
- Lemon slice

Combine all liquid ingredients in a cocktail shaker with cracked ice. Shake well and strain into an Old-Fashioned glass and garnish with lemon slice.

CUCARACHA

- 1 part gold tequila ($^1/_2$ oz.)
- 1 part Kahlúa ($^1/_2$ oz.)
- 1 part Coca-Cola® ($^1/_2$ oz.)

Combine all the ingredients in a shot glass and drink right away.

DOUBLE EAGLE

- 3 parts white tequila (1$^1/_2$ oz.)
- 3 parts Cointreau (1$^1/_2$ oz.)

Pour ingredients over ice cubes into a brandy snifter.

FIRE ALARM

- 2 parts white tequila (1 oz.)
- Tabasco® (to taste)

Pour tequila into a chilled shot glass and top with Tabasco.

FIRE AND ICE

- 1 part white tequila ($^1/_2$ oz.)
- 1 part peppermint schnapps ($^1/_2$ oz.)

Stir the ingredients in a mixing glass with cracked ice. Strain into a chilled shot glass.

GREEN LIZARD

- 1 part white tequila ($^1/_2$ oz.)
- 2 parts green crème de menthe (1 oz.)

Pour ingredients in a chilled Old-Fashioned glass with ice cubes and stir.

GRENADE

- 6 parts cranberry juice (3 oz.)
- 2 parts white tequila (1 oz.)
- Orange peel for garnish

Combine liquid ingredients in a mixing glass with ice cubes and stir. Strain into a chilled cocktail shaker and garnish with orange peel.

HONEYCOMB

- 4 parts gold tequila (2 oz.)
- 3 parts whisky sour mix (1$^1/_2$ oz.)
- 2 teaspoons honey

Combine all ingredients in a cocktail shaker with cracked ice and shake well. Strain into a chilled collins glass over ice cubes.

ICED TEQUILA

- 3 parts gold tequila (1$^1/_2$ oz.)
- 1 part iced tea ($^1/_2$ oz.)

Combine ingredients in a cocktail shaker with cracked ice and shake well. Strain into a chilled shot glass.

LA FRESCA

- 4 parts Sauza Hornitos (2 oz.)
- Squirt®
- Coarse salt
- Juice of 2 limes

Place a pinch of salt in the bottom of a highball glass. Add ice cubes, the juice of 2 limes and Sauza Hornitos. Then fill the glass with Squirt® and stir.

LA FRESCA

MEXICAN COCKTAIL

- 2 parts gold tequila (1 oz.)
- 4 parts Champagne (2 oz.)
- 1 part lime juice ($^1/_2$ oz.)
- Sugar (to taste)

Pour all ingredients in a mixing glass, stir gently and pour into a chilled champagne glass.

MEXICAN SLOE SCREW

- 3 parts white tequila ($1^1/_2$ oz.)
- 6 parts orange juice (3 oz.)
- 1 part sloe gin ($^1/_2$ oz.)
- Lime wedge

Pour tequila and orange juice into a highball glass over ice cubes and stir. Top with gin and garnish with lime.

MESA CITY PRICKLY PEAR MARGARITA

- 4 parts white tequila (2 oz.)
- 2 parts Cointreau (1 oz.)
- 2 parts prickly pear juice (1 oz.)
- 1 part lime juice ($^1/_2$ oz.)
- Coarse salt
- Lime wedge

Rub the lime wedge around the rim of a cocktail glass and dip the rim into a saucer of coarse salt. Combine the remaining ingredients in a cocktail shaker with ice cubes and shake well. Strain into the chilled, salted cocktail glass.

MEXICAN PAIN KILLER

- 1 part gold tequila ($^1/_2$ oz.)
- 1 part vodka ($^1/_2$ oz.)
- 1 part light rum ($^1/_2$ oz.)
- 2 parts pineapple juice (1 oz.)
- 1 part orange juice ($^1/_2$ oz.)
- 2 tablespoons cream of coconut*

Combine the ingredients with cracked ice in a blender. Blend until smooth and pour into a collins glass.

*Coco Lopez® recommended

NATIVE SUN

- 2 parts gold tequila (1 oz.)
- 2 parts amaretto (1 oz.)
- Orange twist

Combine liquid ingredients in a cocktail shaker with cracked ice and shake well. Strain into a chilled cocktail glass and garnish with orange twist.

MESA CITY PRICKLY PEAR MARGARITA

PIMBI

- 3 parts white tequila ($1^1/_2$ oz.)
- 6 parts pineapple juice (3 oz.)
- 2 parts lime juice (2 oz.)
- 1 teaspoon sugar

Combine all ingredients in a cocktail shaker with cracked ice and shake well. Strain into a chilled highball glass over ice cubes.

PINEAPPLE SUNRISE

- 3 parts white tequila ($1\,^1/_2$ oz.)
- 6 parts pineapple juice (3 oz.)
- 1 part lime juice ($^1/_2$ oz.)
- 1 teaspoon grenadine
- Slice of fresh pineapple

Combine liquid ingredients in a cocktail shaker with cracked ice and shake well. Strain into a chilled cocktail glass and garnish with pineapple slice.

SILK STOCKING

- 4 parts crème de cacao (2 oz.)
- 1 can evaporated milk ($3^1/_2$ oz.)
- 4 parts grenadine (2 oz.)
- 4 parts white tequila (2 oz.)
- Cinnamon

Combine liquid ingredients in a blender with ice. Blend at medium speed until smooth. Pour into a chilled highball glass and sprinkle with cinnamon.

SUMMER BREEZE

- 4 parts white tequila (2 oz.)
- 4 parts soda water (2 oz.)
- Mint leaves (to taste)
- 1 teaspoon sugar
- 1 tablespoon hot water
- 1 lime

Muddle the mint leaves, sugar and hot water in the bottom of a mixing glass and fill with ice cubes. Pour in tequila, soda water and a sqeeze of lime and stir vigorously. Strain into a chilled cocktail glass and garnish with mint leaf.

SWEET AND SOUR

- 2 parts white tequila (1 oz.)
- 1 tablespoon lemon juice
- 1 part grenadine ($^1/_2$ oz.)

Combine tequila and lemon juice in a cocktail shaker with cracked ice and shake well. Strain into a chilled shot glass and carefully top with grenadine.

SWEET AND SOUR

TEQUILA COCKTAIL

- 6 parts gold tequila (3 oz.)
- 2 parts lime juice (1 oz.)
- $^1/_4$ teaspoon grenadine
- Dash of Angostura bitters

Combine all ingredients in a cocktail shaker with cracked ice. Shake well and strain into a chilled cocktail glass.

TEQUILA COLLINS

- 4 parts white tequila (2 oz.)
- 2 parts lemon juice (1 oz.)
- 1 teaspoon sugar syrup
- Sparkling water
- Maraschino cherry

Pour tequila into a chilled collins glass over ice cubes. Add lemon juice and syrup. Stir well and add sparkling water. Stir gently and garnish with cherry.

TEQUILA FIZZ

- 6 parts white tequila (3 oz.)
- 2 parts lime juice (1 oz.)
- 2 parts grenadine (1 oz.)
- 1 egg white
- Ginger ale

Combine all ingredients except ginger ale with cracked ice in a cocktail shaker. Shake vigorously and strain into a chilled Collins glass over ice cubes. Fill with ginger ale and stir gently.

TEQUILA GHOST

- 4 parts white tequila (2 oz.)
- 2 parts pernod (1 oz.)
- 1 part lemon juice ($^1/_2$ oz.)

Combine all ingredients with cracked ice in a cocktail shaker and shake well. Strain into a chilled cocktail glass.

TEQUILA COLLINS

TEQUILA GIMLET

- 6 parts white tequila (3 oz.)
- 2 parts Rose's® lime juice (1 oz.)
- Lime slice

Pour tequila and lime juice into an Old-Fashioned glass filled with ice cubes. Stir and garnish with lime wedge.

TEQUILA MANHATTAN

- 6 parts gold tequila (3 oz.)
- 2 parts sweet vermouth (1 oz.)
- 1 teaspoon lime juice
- Maraschino cherry
- Orange slice

Combine liquid ingredients in a cocktail shaker with cracked ice and shake well. Strain into a chilled Old-Fashioned glass over ice cubes. Garnish with cherry and orange slice.

TEQUILA MARIA

- 4 parts white tequila (2 oz.)
- 8 parts tomato juice (4 oz.)
- 1 part lime juice ($^1/_2$ oz.)
- 1 teaspoon white horseradish
- Tabasco sauce to taste
- 3–5 dashes Worcestershire sauce
- Ground black pepper to taste
- Celery salt to taste
- Pinch of cilantro
- Lime wedge

Combine all ingredients except lime wedge in a mixing glass with cracked ice. Pour into a chilled Old-Fashioned glass and garnish with lime wedge.

TEQUILA MOCKINGBIRD

- 4 parts white tequila (2 oz.)
- 2 parts white crème de menthe (1 oz.)
- 2 parts lime juice (1 oz.)

Combine all ingredients in a cocktail shaker with cracked ice and shake well. Strain into a chilled cocktail glass.

TEQUILA MOCKINGBIRD

TEQUILA OLD-FASHIONED

- 4 parts gold tequila (2 oz.)
- 1 teaspoon sugar
- 3–5 dashes Angostura bitters
- Sparkling water
- Maraschino cherry

Combine sugar, water and bitters in the bottom of a chilled Old-Fashioned glass. Fill with ice cubes and add tequila. Stir well and garnish with cherry.

TEQUILA SHOT

- 4 parts tequila of your choice (2 oz.)
- Lemon wedge
- Salt

Pour tequila into a shot glass. Moisten your hand between thumb and forefinger and put salt on it. Lick the salt, down the shot and suck on the lemon.

TEQUILA SOUR

TEQUILA SOUR

- 4 parts white tequila (2 oz.)
- 3 parts lemon juice ($1^1/_2$ oz.)
- 1 teaspoon sugar
- Lemon slice
- Maraschino cherry

Combine all ingredients except fruit in a cocktail shaker with cracked ice and shake well. Strain into a chilled sour glass and garnish with fruit.

TEQUILA SPLASH

- 3 parts gold tequila ($1^1/_2$ oz.)
- Orange juice

Pour tequila into an Old-Fashioned glass over ice cubes. Fill with orange juice and stir.

TEQUILA STINGER

- 4 parts gold tequila (2 oz.)
- 2 parts white crème de menthe (1 oz.)

Combine ingredients in a cocktail shaker with cracked ice and shake well. Strain into a chilled cocktail glass.

TEQUILA SUNRISE

- 4 parts white tequila (2 oz.)
- Orange juice
- 2 parts grenadine (1 oz.)

Pour tequila into a chilled highball glass over ice cubes. Fill glass with orange juice, leaving a little room on top and stir. Slowly pour in grenadine.

TEQUILA SUNRISE

TEQUILADA

- 4 parts gold tequila (2 oz.)
- 4 parts cream of coconut (2 oz.)
- 8 parts pineapple juice (4 oz.)
- Slice of fresh pineapple

Combine the ingredients except for the pineapple in a cocktail shaker with cracked ice and shake well. Strain into a chilled cocktail glass and garnish with pineapple slice.

TEQUINI

- 6 parts white tequila (3 oz.)
- 1 part dry vermouth ($^1/_2$ oz.)
- Dash of Angostura bitters
- Lemon twist

Stir all liquid ingredients in a mixing glass with cracked ice. Strain into a chilled cocktail glass and garnish with lemon twist.

TEQUONIC

- 4 parts silver tequila (2 oz.)
- 3 parts fresh lime juice ($1^1/_2$ oz.)
- Tonic water
- Lime wedge

Pour tequila into a chilled highball glass over ice cubes. Add juice and stir. Fill with tonic and garnish with lime wedge.

TUN-TUN

- 3 parts white tequila ($1^1/_2$ oz.)
- 2 teaspoons lemon juice
- 1 tablespoon raspberry syrup
- Dash of Grand Marnier
- Orange slice and strawberry

Combine all ingredients except Grand Marnier and fruit in a cocktail shaker and shake well. Pour into a cocktail glass over shaved ice and stir. Float Grand Marnier on top and garnish with fruit.

GAZPACHO WITH TEQUILA-PICKLED ONION

SERVES SIX

GAZPACHO:

4 slices French or Italian bread, crusts removed, torn into chunks
12 ounces ripe red tomatoes, coarsely chopped
1 medium red onion, coarsely chopped
2 cloves garlic, coarsely chopped
1 large cucumber, coarsely chopped
2 medium red bell peppers, coarsely chopped
2 tablespoons Spanish olive oil
2 tablespoons sherry vinegar
1 tablespoon tequila
Salt and freshly ground pepper

TEQUILA-PICKLED ONIONS:

2 large red onions, cut in half and thinly sliced
1 cup white wine vinegar
$^1/_4$ cup tequila
1 teaspoon granulated sugar
2 cloves garlic, finely chopped
1 tablespoon finely chopped thyme
Salt and freshly ground pepper

GARNISH: $^1/_4$ cup finely chopped cilantro

TO MAKE THE ONIONS:

1. In a small saucepan combine the vinegar, tequila, sugar and garlic. Over medium heat, bring to a boil and simmer for a few minutes.

2. In a medium bowl, place the onions, pour in the mixture, add the thyme and season with salt and pepper to taste. Let sit at room temperature for at least 2 hours.
Drain the onions.

TO MAKE THE GAZPACHO:

1. Pour water over bread to cover, squeeze out the water and place bread in the bowl of a food processor. Add the remaining ingredients to the food processor and puree until smooth. (If too thick, add a little ice water to thin out.)

2. Strain through a fine sieve and season with salt and pepper to taste.

3. Cover and refrigerate for at least 2 hours. Divide among six bowls and garnish with the pickled onions and cilantro.

BOBBY FLAY'S TEQUILA-CURED SALMON

SERVES SIX

1 salmon filet, skin on (about 2 pounds)
1 tablespoon canned puréed chipotles
$^1/_4$ cup chopped cilantro
$^1/_4$ cup whole mustard seeds
2 tablespoons cumin seeds
Zest of 5 limes, in strips

2 cups kosher salt
3 cups (packed) light brown sugar
1 cup tequila

1. Place salmon filet on a baking sheet, flesh side up. Spread with puréed chipotles. Sprinkle evenly with cilantro, mustard seeds, cumin seeds and lime zest.

2. In a mixing bowl combine the salt, brown sugar and tequila. Cover the salmon with the salt mixture.

3. Place another baking sheet on top of the salmon and weigh it down with a cutting board or similar weight. Refrigerate for 48 hours.

4. Remove the salmon from the refrigerator and scrape off the curing ingredients. When cured, the fish should have a raw appearance and a firm, but not hard, texture. If the color changes to pale pink, the curing has gone on for too long, and the fish is "cooked."

BELGIAN ENDIVE-JÍCAMA SALAD WITH TANGERINE-TEQUILA VINAIGRETTE

SERVES SIX

VINAIGRETTE:

1/4 cup fresh tangerine juice
2 tablespoons tequila
1 tablespoon white wine vinegar

1 small shallot, coarsely chopped
$^{3}/_{4}$ cup Spanish olive oil
Salt and freshly ground white pepper to taste

SALAD:

2 heads Belgian endive
4 tangerines, peeled and sectioned
1 medium jícama, peeled and julienned
Salt and freshly ground white pepper to taste
$^{1}/_{2}$ cup finely chopped cilantro

TO MAKE THE VINAIGRETTE:

Place the juice, tequila, vinegar and shallot in a blender and blend until smooth. With the motor running, slowly add the olive oil until completely blended. Season with salt and white pepper.

TO MAKE THE SALAD:

Arrange the endive around the edge of each plate. Toss the tangerine sections and jícama with $^{1}/_{4}$ cup of the vinaigrette and season lightly with salt and pepper. Mound the mixture in the middle of each plate, drizzle with additional vinaigrette and sprinkle with cilantro.

TEQUILA-AND-LIME-STEAMED MUSSELS

SERVES FOUR

1 tablespoon olive oil
1 tablespoon freshly grated ginger
1 tablespoon finely chopped garlic
$\frac{1}{4}$ cup finely chopped red onion
1 cup tequila
1 tablespoon finely grated lime zest
32 mussels, cleaned, scrubbed and debearded
Salt and freshly ground pepper to taste
2 tablespoons finely chopped cilantro

Heat the oil in a medium stockpot over medium heat. Add the ginger, garlic and onion and cook until onion is translucent. Raise the heat to high and add the tequila, salt and pepper and bring to a boil. Add the mussels and cilantro and stir well to blend. Cover the pot and cook until the mussels open, about 3 to 4 minutes. Remove the mussels to a large bowl and serve immediately.

TEQUILA-MARINATED GRILLED VEGETABLE SALAD

SERVES FOUR

$\frac{1}{4}$ cup olive oil
2 tablespoons tequila
1 tablespoon finely chopped fresh thyme
1 tablespoon finely chopped fresh flat-leaf parsley
1 tablespoon finely chopped garlic
2 large red peppers, quartered and seeds removed
2 large yellow peppers, quartered and seeds removed
2 small zucchini, quartered
12 button mushrooms
12 cherry tomatoes
Salt and freshly ground pepper to taste

1. Combine the oil, tequila and herbs in a large shallow baking dish. Add the vegetables and toss to coat. Let sit at room temperature for at least 1 hour.

2. Preheat the grill or broiler. Remove the vegetables from the marinade and season with salt and pepper. Grill the vegetables on both sides until just cooked through.

OYSTERS ON THE HALF SHELL WITH TEQUILA MIGNONETTE SAUCE

SERVES FOUR

3/4 cup tequila
3 shallots, minced
Pinch of sugar
Salt and coarsely ground fresh pepper to taste
24 raw oysters, scrubbed well

1. For the mignonette sauce: In a small bowl, whisk together the tequila, shallots, sugar and salt and pepper. Set aside.

2. Shuck the oysters, keeping the round bottom shells and discarding the top shells. Place an oyster in each of the reserved shells and arrange on a serving platter that has been lined with crushed ice. Spoon on the mignonette sauce and serve immediately.

CHILLED LOBSTER SALAD WITH TEQUILA-CILANTRO VINAIGRETTE

SERVES FOUR

VINAIGRETTE:

3/4 cup olive oil
3 tablespoons fresh lime juice
1 tablespoon tequila
1/4 cup chopped cilantro
1 teaspoon honey
Salt and freshly ground pepper to taste

SALAD:

4 cups mesclun greens
2 small tomatoes, quartered
2 hard cooked eggs, quartered
1 red bell pepper, julienned
4 one-pound lobsters, steamed and meat removed

1. To make the vinaigrette, place all ingredients in a blender and blend until smooth. Season with salt and pepper.

2. To make the salad, divide greens among four large plates. Decoratively arrange tomatoes, eggs, peppers and lobster on each plate. Drizzle lightly with the vinaigrette.

SHRIMP COCKTAIL WITH FRESH TOMATO-TEQUILA COCKTAIL SAUCE

SERVES FOUR

COCKTAIL SAUCE:

5 ripe plum tomatoes, quartered and seeds removed
2 tablespoons ketchup
2 tablespoons prepared horseradish in vinegar
1 tablespoon tequila
Pinch of sugar
Dash of hot sauce
Salt and freshly ground pepper to taste

SHRIMP:

2 quarts water
2 tablespoons coarse salt
1 lemon, halved
1 pound large shrimp, peeled and deveined
(about 2 dozen)

1. To make the cocktail sauce, place all ingredients in the bowl of a food processor and pulse until the tomatoes are still slightly chunky. Season with salt and pepper to taste. Refrigerate, covered, for at least 1 hour.

2. To cook the shirmp: In a large heavy saucepan, place the water, salt and lemon and bring to a boil. Add the shrimp, turn off the heat and stir. Test the shrimp for doneness after

2 to 3 minutes. Remove when completely cooked through and drain. Refrigerate, covered, for at least 1 hour. Serve with cocktail sauce.

DRUNKEN BEANS

SERVES FOUR

1/4 cup olive oil
1 large red onion, halved and thinly sliced
2 cloves garlic, finely chopped
2 plum tomatoes, cored, seeded and finely chopped
2 serrano chiles, seeded and finely chopped
1/4 cup finely chopped cilantro
1/4 pound pinto beans, cooked
(or 3 cups canned beans, rinsed)
Salt to taste
3/4 cup tequila

1. In a saucepan heat oil over medium-high heat. Sauté onions until lightly browned. Stir in tomatoes, chiles and cilantro and cook for 1 minute.

2. Add the cooked beans, salt and tequila.

3. Cook uncovered over low heat until the juices have thickened, about 30 minutes.

TEQUILA CEVICHE

SERVES FOUR

1/2 pound medium shrimp, shelled, deveined, parboiled and cut into 1/2-inch dice
1/2 pound sea scallops, cut into 1/2-inch dice
1/2 pound salmon filet, cut into 1/2-inch dice
1/2 cup diced tomatoes
1/2 cup diced mango
1/4 cup diced red onion
1 jalapeño pepper, minced
2 cups fresh lime juice
1/2 cup tequila
1/2 cup coarsely chopped cilantro

1 tablespoon sugar
Salt and freshly ground pepper to taste
1 grapefruit, peeled and segmented
1 orange, peeled and segmented

1. In a large nonreactive (i.e., glass or ceramic) mixing bowl, combine the shrimp, scallops, salmon, tomatoes, mango, jalapeño, lime juice and tequila. Cover and marinate in the refrigerator for 2 1/2 to 3 hours, stirring mixture after 1 hour.

2. Just before serving, drain off as much lime-tequila juice as possible and add the cilantro, sugar and salt and pepper. Gently fold in the grapefruit and orange segments.

TEQUILA-SCENTED SALSA VERDE

SERVES FOUR

10 small tomatillos, husks removed, washed well and quartered
2 cloves garlic, finely chopped
1 teaspoon honey
2 tablespoons finely minced scallion, white part only
1 tablespoon red onion
1 small jalapeño, finely chopped
3 tablespoons coarsely chopped fresh cilantro
1 tablespoon tequila
Salt and freshly ground pepper to taste
Pinch of sugar

1. Place the tomatillos, garlic and honey in the bowl of a food processor and pulse until mixture is slightly chunky. Transfer to a medium bowl.

2. Add the scallions, onion, jalapeño, cilantro and tequila and mix well. Let the salsa sit at room temperature for 15 minutes. Season with salt and pepper, adding more sugar if the salsa is too acidic. Serve with tortilla chips.

FISH AND SEAFOOD

PINEAPPLE-TEQUILA-GLAZED TUNA STEAKS

SERVES FOUR

3 cups pineapple juice
2 tablespoons tequila
4 six-ounce tuna steaks
Salt and freshly ground pepper to taste

1. In a small, nonreactive saucepan, cook the pineapple juice over medium-high heat until reduced to $^1/_2$ cup. Remove from the heat and add the tequila. Let cool.

2. In a medium-sized shallow baking dish place the tuna steaks and pour in reduced glaze. Turn steaks over to coat completely. Cover and refrigerate for 1 hour.

3. Heat a grill pan over high heat until almost smoking. Season tuna steaks with salt and pepper and grill on one side for 2 to 3 minutes, or until golden brown. Turn steaks over and continue cooking for 1 to 2 minutes for rare doneness.

GRILLED SWORDFISH WITH TEQUILA-BLACK PEPPER BUTTER

SERVES FOUR

$^1/_4$ cup tequila
2 tablespoons fresh orange juice
$^1/_2$ pound (2 sticks) unsalted butter, softened
$^1/_4$ teaspoon salt
1 teaspoon coarsely ground fresh black pepper
4 six-ounce swordfish steaks
2 tablespoons olive oil
Salt and freshly ground black pepper to taste

1. To make the butter, in a small, nonreactive saucepan place the tequila and orange juice and reduce to 2 tablespoons. Let cool.

2. In a medium bowl place the softened butter. Add the reduced tequila-orange liquid, salt and pepper and mix until combined. Cover and chill in the refrigerator until solid, about 2 hours.

3. Brush the swordfish with olive oil and season both sides with salt and pepper. Heat a grill pan over high heat until almost smoking. Grill the swordfish on one side for 3 to 4 minutes, or until golden brown. Turn over and continue cooking for 3 to 4 minutes for medium doneness. Serve immediately with a dollop of the tequila-black pepper butter.

RED SNAPPER POACHED IN SPICY TEQUILA MARINADE

SERVES FOUR

POACHING LIQUID:

4 cups fish stock
½ cup tequila
2 serrano chile peppers, cut in half
1 small red onion, coarsely chopped
6 sprigs flat-leaf parsley
6 whole black peppercorns
1 bay leaf
4 six-ounce red snapper filets
Salt and freshly ground pepper to taste

1. In a medium skillet place the poaching liquid ingredients. Bring to a boil for 2 minutes and reduce the heat to a simmer.

2. Season the snapper filets on both sides with salt and pepper. Place the filets into the poaching liquid and cook to medium-well doneness, about 6 to 8 minutes.

GRILLED TEQUILA-SOAKED SHRIMP WITH TOMATO RELISH

SERVES FOUR

1 cup tequila
3 cloves garlic, coarsely chopped
$^1/_4$ cup olive oil
2 tablespoons fresh lime juice
1 teaspoon sugar
Salt and freshly ground pepper to taste
24 large shrimp, peeled

TOMATO RELISH:

6 plum tomatoes, seeded and coarsely chopped
2 cloves garlic, finely chopped
2 tablespoons olive oil
2 tablespoons basil chiffonade
Salt and freshly ground pepper to taste

1. To make the tomato relish, in a medium bowl combine all of the ingredients and let sit at room temperature for 30 minutes.

2. To make the shrimp, combine the tequila, garlic, olive oil, lime juice and sugar in a medium shallow baking dish. Add the shrimp and toss to coat. Let sit at room temperature for 20 minutes. Heat the grill. Remove the shrimp from the marinade and season with salt and pepper. Grill for 2 to 3 minutes on each side, or until cooked through.

SALMON BAKED IN TEQUILA SALSA VERDE

SERVES SIX

SALSA VERDE:

4 tablespoons olive oil, divided
7 tomatillos, husks removed
3 tablespoons finely chopped parsley

3 tablespoons finely chopped basil
2 tablespoons finely chopped tarragon
$1/2$ cup water
4 cloves garlic, finely chopped
$1/2$ cup tequila
6 cups clam juice
Juice of 1 lemon
Salt and freshly ground pepper to taste

SALMON:

6 salmon filets (5 to 6 ounces each)
2 tablespoons olive oil
Salt and freshly ground pepper to taste

1. To make the salsa verde: In a large sauté pan, heat 2 tablespoons of the olive oil over high heat until almost smoking. Add the tomatillos, sauté for 3 to 4 minutes, remove from the pan and roughly chop them.

2. Place the tomatillos in a food processor, add the parsley, basil, tarragon and water and purée until smooth.

3. In a medium saucepan, place the remaining 2 tablespoons of olive oil. Over medium heat sauté the garlic until soft. Raise the heat to medium-high, add the tequila and cook until most of the liquid has evaporated. Add the clam juice and reduce to $1/2$ cup, about 15 minutes. Add the lemon juice, tomatillo purée and season to taste with salt and pepper. Set aside.

4. To make the salmon: Preheat the oven to 350° Farenheit. In a Dutch oven, heat the olive oil over medium-high heat. Season the salmon on both sides, place the skin side down in the oil and cook for 2 to 3 minutes, or until golden. Turn the salmon filets over, drain off the excess oil and cover with the salsa verde. Cook covered for 10 to 12 minutes, or until done.

GRILLED SEA SCALLOPS

SERVES SIX

SAUCE:

2 tablespoons unsalted butter
1 large onion, finely chopped
3 cloves garlic, finely chopped
1/4 cup tequila
1 cup clam juice
3 cups unsweetened coconut milk
1 large poblano pepper, roasted
1/4 cup cilantro
1/2 cup spinach leaves, washed
Salt and freshly ground pepper to taste

SCALLOPS:

24 fresh sea scallops, washed and dried
2 tablespoons olive oil
Salt and freshly ground pepper to taste
Toasted coconut, for garnish
Chopped cilantro, for garnish

1. To make the sauce, in a medium saucepan melt the butter over medium-high heat. Add the onion and garlic and cook until soft.

2. Add the tequila, raise the heat and cook until tequila is mostly evaporated. Add the clam juice and reduce by half. Add the coconut milk and reduce by half.

3. Add the pepper, cilantro and spinach and cook for 2 minutes.

4. Preheat the grill or heat a grill pan over high-heat until almost smoking. Brush both sides of the scallops with olive oil and season with salt and pepper.

5. Grill the scallops for 2 minutes on one side, turn over, and cook an additional 2 to 3 minutes.

6. Divide the scallops among the plates and drizzle with the sauce. Garnish with the toasted coconut and chopped cilantro.

POULTRY

TEQUILA HOT WINGS WITH BLUE CHEESE DIP

SERVES FOUR

BLUE CHEESE DIP:

1/2 cup sour cream
1/2 cup mayonnaise
1 teaspoon tequila
1 tablespoon finely chopped cilantro
2 tablespoons finely chopped green onion
2 cloves garlic, finely chopped

1 teaspoon Tabasco® sauce
3 tablespoons crumbled blue cheese
Salt and freshly ground pepper to taste

CHICKEN WINGS:

24 chicken wings
2 cups peanut oil
1 stick unsalted butter
1 tablespoon Tabasco® sauce
1 tablespoon tequila

JÍCAMA STICKS:

1. For the dip, in a medium bowl combine all the ingredients. Cover and refrigerate for at least 1 hour.

2. Rinse the wings and pat dry with paper towels. In a medium saucepan, heat the oil to 350° Farenheit. Season the wings with salt and pepper and fry in batches for about 5 minutes, or until golden brown on all sides. Remove wings and drain on paper towels.

3. In a medium sauté pan, heat the butter until melted. Add the Tabasco and tequila and stir to combine. Add the wings to the sauce and toss to coat completely. Serve with dip and jícama sticks.

GRILLED DUCK BREAST WITH APRICOT-TEQUILA GLAZE

SERVES FOUR

APRICOT-TEQUILA GLAZE:

1 cup apricot preserves
1 teaspoon grated fresh ginger
1 teaspoon Dijon mustard
1 tablespoon tequila

DUCK:

2 whole duck breasts, bones in and skin on, trimmed of excess fat
1 tablespoon olive oil
Salt and freshly ground pepper to taste

1. To make the glaze, mix all the ingredients in a saucepan and heat to melt the preserves. Let cool.

2. To make the duck, heat a grill to medium-high. Brush the skin side with the apricot-tequila glaze and season with salt and pepper. Place the breasts skin-side down on the grill and cook for 4 to 5 minutes, or until golden brown. Turn the breasts over, brush with the glaze, season with salt and pepper and continue cooking an additional 4 to 5 minutes for medium doneness. Remove from the grill and let rest 5 minutes before slicing.

ROASTED TURKEY BREAST WITH CRANBERRY-TEQUILA RELISH

SERVES SIX

TURKEY:

One 4- to 5-pound turkey breast
1 tablespoon olive oil
Salt and freshly ground pepper to taste

CRANBERRY-TEQUILA RELISH:

2 tablespoons unsalted butter
4 cloves garlic, finely chopped
1 small Spanish onion, finely chopped
2 tablespoons tequila
1 jalapeño pepper, seeded and finely chopped

1 pound fresh cranberries, washed
2 cups orange juice
$\frac{1}{2}$ cup brown sugar
Salt and freshly ground pepper to taste

1. Preheat the oven to 350° Farenheit. Rub the turkey with olive oil and season with salt and pepper. In a roasting pan, roast the turkey until the juices run clear, approximately 1 to 1 $\frac{1}{2}$ hours. Let sit for 15 minutes before slicing.

2. Meanwhile, make the cranberry-tequila relish. In a large saucepan, melt the butter over medium-high heat. Add the garlic and onion and cook until soft. Add the tequila and cook for 2 minutes. Add the jalapeño, cranberries, orange juice, brown sugar and season with salt and pepper. Reduce heat to medium-low and cook for 10 to 15 minutes, or until the berries pop. Transfer to a bowl, let cool to room temperature and serve with the turkey.

MEAT

TEQUILA-LIME MARINATED FLANK STEAK

SERVES SIX

MARINADE:

3 tablespoons Spanish olive oil
3 tablespoons tequila
2 tablespoons fresh lime juice
2 dashes Tabasco® sauce
4 cloves garlic, coarsely chopped
1 small Spanish onion, coarsely chopped

FLANK STEAK:

One 1 $\frac{1}{2}$-pound flank steak
Salt and freshly ground pepper to taste
12 flour tortillas
Pico de Gallo
Guacamole
Cilantro leaves, for garnish

1. To make the marinade: In a large bowl combine the ingredients. Add the flank steak and toss to coat completely with the marinade. Cover and place in the refrigerator for 2 hours, turning the meat after 1 hour.

2. To make the flank steak: 30 minutes before you are ready to cook, heat the grill. Remove the steak from the marinade and pat dry with a paper towel. Season both sides of the steak with salt and pepper. Grill for 3 to 4 minutes per side for medium-rare doneness. Let rest for 10 minutes before slicing.

3. Slice the steak against the grain into 1/4-inch thick diagonal strips. Serve rolled in warm tortillas with salsa, guacamole and cilantro.

DESSERTS

TEQUILA-LIME TART

SERVES EIGHT

TART CRUST:

1 1/2 cups all-purpose flour
1/4 teaspoon salt
1/4 cup granulated sugar
1/2 stick unsalted butter, chilled and cut into tablespoons

1 egg yolk
1 $1/2$ tablespoons cold water

TEQUILA-LIME FILLING:

5 large egg yolks
20 ounces sweetened condensed milk
$1/2$ cup fresh or bottled Key lime juice
2 tablespoons tequila
1 teaspoon finely chopped lime zest

FRESHLY WHIPPED SWEETENED CREAM, FOR GARNISH

1. Preheat the oven to 350° Farenheit. Place the flour and butter in the bowl of a food processor and pulse until mixture resembles coarse grain. With the machine running, slowly add the egg yolk and 1 tablespoon of water. Continue processing until the dough forms a ball, adding more water if necessary.

2. Press the dough evenly into a 9-inch tart pan with removable bottom. With a fork, poke holes in the dough and bake for 20 to 25 minutes, or until light golden brown.

3. To make the filling, in a large bowl whisk the egg yolks and condensed milk until just combined.

4. Pour the filling into prebaked shell and bake just until center is firm, about 9 to 10 minutes.

5. Chill in the refrigerator completely before removing from the pan. Slice and serve with a dollop of whipped cream.

TEQUILA CHEESECAKE WITH TEQUILA-LIME GLAZE

YIELDS ONE 8-INCH CHEESECAKE

CRUST:

$1/2$ cup shelled pumpkin seeds, lightly toasted
1 cup graham cracker crumbs
2 tablespoons granulated sugar
6 tablespoons unsalted butter, melted

CAKE:

1 $^{1}/_{2}$ pounds cream cheese, at room temperature
6 tablespoons unsalted butter, at room temperature
1 cup granulated sugar
2 tablespoons cornstarch
4 large eggs, at room temperature
1 cup sour cream
$^{1}/_{4}$ cup tequila
2 teaspoons vanilla extract

GLAZE:

1 cup lime marmalade
2 tablespoons Rose's® lime juice
1 tablespoon tequila
1 tablespoon water

1. Preheat the oven to 300° Farenheit. Grease an 8-inch cake pan or springform pan with melted butter.

2. In a large bowl combine the pumpkin seeds, graham cracker crumbs, sugar and melted butter and blend well. Pat the mixture evenly over the bottom of the prepared pan.

3. In a mixing bowl place the cream cheese, butter, sugar and cornstarch and beat at medium speed just to blend. Add the eggs one at a time until each is incorporated. Reduce the speed and beat in the sour cream, tequila and vanilla.

4. Pour the batter into the prepared pan. Set the pan in a larger pan such as a roasting pan, and pour about 1 inch of hot water into the larger pan. (If using a springform pan, wrap the bottom with aluminum foil.)

5. Bake for 1 $^1/_4$ to 1 $^1/_2$ hours, or until lightly tanned, slightly puffed and barely firm.

6. Cool to room temperature in the water bath. Remove from larger pan, refrigerate overnight and remove from cake or springform pan.

7. To make the glaze, place all the ingredients in a small saucepan and cook over medium heat until smooth, stirring constantly. Let cool for 5 minutes and pour over the top of the cheesecake.

DRUNKEN STRAWBERRY SALSA

YIELDS 1 ½ CUPS

1 cup ripe strawberries, hulled and quartered

¼ teaspoon minced serrano or jalapeño chile, seeds removed

1 tablespoon chopped fresh mint

1 tablespoon gold tequila

1 teaspoon lime juice

2 tablespoons sugar

1. In a medium bowl combine half the strawberries and the remaining ingredients. Refrigerate for 30 minutes to 3 hours, tossing occasionally. Just before serving, stir in the remaining strawberries.

Serving suggestions: Serve over ice cream or pound cake.

STRAWBERRY MARGARITA SORBET

SERVES SIX

4 cups ripe strawberries, hulled and sliced in half
3/4 cup simple syrup (equal parts sugar and water)
Juice of 1 lime
1/2 cup of tequila
1/4 cup triple sec or any orange-flavored liqueur
6 whole strawberries, for garnish
Mint sprigs, for garnish

1. Place all the ingredients except the garnish in a blender and blend until smooth. Taste the mixture for sweetness and

add more simple syrup if needed.

2. Strain through a fine sieve and refrigerate until cold.

3. Place in an ice cream machine and process according to the manufacturer's instructions. Freeze until firm.

PUBLIC LIBRARY
MENTOR, OH